Connectionist Robot Motion Planning

A Neurally-Inspired Approach to Visually-Guided Reaching

Perspectives in Artificial Intelligence *Volume 7*

Connectionist Robot Motion Planning

A Neurally-Inspired Approach to Visually-Guided Reaching

Bartlett W. Mel

Computation and Neural Systems Program
California Institute of Technology
Pasadena, California

ACADEMIC PRESS, INC.
Harcourt Brace Jovanovich, Publishers
Boston San Diego New York
London Sydney Tokyo Toronto

This book is printed on acid-free paper. ♾

ACADEMIC PRESS, INC.
1250 Sixth Avenue, San Diego, CA 92101

United Kingdom Edition published by
ACADEMIC PRESS LIMITED
24-28 Oval Road, London NW1 7DX

Library of Congress Cataloging-in-Publication Data
Mel, Bartlett W.
Connectionist robot motion planning : a neurally-inspired approach to visually-guided reaching / Bartlett W. Mel.
p. cm. — (Perspectives in artificial intelligence : v. 7)
Includes bibliographical references.
Includes index.
CN - ISBN 0-12-490020-8 (alk. paper)
1. Robots — Motion. 2. Artificial intelligence. I. Title.
II. Series.
TJ211.4.M43 1990
629.8'92 — dc20 90-40669
CIP

Printed in the United States of America
90 91 92 93 9 8 7 6 5 4 3 2 1

To my mother and father,
and
to Maria

Editor's Note

This is the third in a cluster of books on robotics and related areas that we are publishing as part of the Perspectives in Artificial Intelligence Series: Jon Connell's *Minimalist Mobile Robotics: A Colony-style Architecture for an Artificial Creature*; Randy Beer's *Intelligence as Adaptive Behavior: An Experiment in Computational Neuroethology*; and now Bartlett Mel's *Connectionist Robot Motion Planning: A Neurally-Inspired Approach to Visually-Guided Reaching*. I said in my editorial for the Connell and Beer books that Artificial Intelligence (AI) has been driven largely by the Turing vision of a disembodied, abstract, symbol-processing intelligence. In this view, the processor is connected at one end to sensory systems that supply a symbolic description of the world, and at the other end to motor systems that accept and execute motion commands. Biological bodies and minds are not decomposed and compartmentalized with such ease, however. Evolution has built higher-level thought on top of a highly integrated and distributed sensory and movement system. Thus, much of the generation and control of motion is mediated by perception, and central processors play only a marginal role in this activity, at best.

The importance of these books to AI is not so much in their contributions to robotics technology, which of course are considerable, but in their contributions to an emerging way of looking at how the body, the senses, and intelligence constitute a more integrated whole than has been imagined by the Turing vision of AI. These three books take the biological paradigm seriously, each in its own way, and constitute challenges to the traditional view of how to support a body and its purposeful movements.

As Mel points out, the problem of visual coordination of hand movements has become the paradigm problem for many researchers trying to understand how perception and motor action are integrated in motion control. Mel's work explores a class of connectionist mechanisms that allow appropriate associations to be learned between visual representations and postures and movements of the arm. These learned relationships are then used as the basis for motion planning.

This work impresses me greatly as a model of good AI research: Mel attacks an important problem whose concrete contributions are important for technology, while at the same time making contact with the fundamental issues of the field. I am impressed especially by the role assigned not simply to perception, but to mental imagistic simulations of motion as a substitute for abstract planning. Thus, in one small package, this book brings together a number of important AI issues. I am pleased to present the work as part of the Perspectives in Artificial Intelligence Series.

— B. Chandrasekaran

Contents

Preface xiii

Acknowledgements xv

1 **Introduction** **1**

2 MURPHY's **Organization** **7**

The Physical Setup 7

The Connectionist Architecture 9

Representational Choices 12

MURPHY's Kinematics 16

Traditional Methods 16

Connectionist Kinematics 19

The Sequential Controller 22

3 **How** MURPHY **Learns** **23**

Learning by Doing 23

Abridged History of the Idea 26

Motivating Sigma-Pi Learning 28

Connectionist Supervised Associative Learning 28

A Historical Quandary 30

New Algorithms Are More Powerful, Less Biological . 32

Changing Assumptions 34

Learning Functions with Lookup-Tables 35

Building Receptive Fields with Multiplication 39

Sigma-Pi Learning 41

The Sigma-Pi Unit 41

The Learning Rule 43
An Example 44
Generalization to Novel Inputs 47
K-d Tree Reimplementation 48

4 MURPHY **in Action** **51**
Planning with Gradient Descent 51
Motion Planning with Obstacles 53
Four Visual Routines 53
The Search Procedure 54
Discussion 64
Visual-Feedback Control 65
System Performance and Scaling Behavior 69
Implementation and Performance Notes 69
Scaling MURPHY Up 70

5 **Robotics Issues** **77**
Learning vs. Built-in Models 77
Style of Representation 78
Using the Full Visual Channel 78
Styles of Robot Motion Planning 82
Artificial Potential Fields: Local Methods 82
Geometric Motion Planning: Global Methods 83
Hybrid Methods: Incorporating Heuristic Search . . . 87
MURPHY Uses Direct Heuristic Search 90

6 **Psychological Issues** **93**
Psychoanalyzing MURPHY 94
Development and Plasticity in Limb Control 95
The Necessity of Active Sensory-Motor Learning . . . 95
Developmental Evidence for Two Modes of Control . . 96
Possible Evidence for Local Learning 98
Models, Imagery, Practice, and Stability 100
Building and Using Mental Models 100
Mental Imagery 102
Mental Practice: Learning by Thinking 108
Corollary Discharges and Perceptual Stability 112
Summary of Psychological Issues 114

7 Biological Issues **117**
The Muscle Interface 117
A Cortical Basis for Visual Limb Control 121
A Programmed-Feedforward Neural Subsystem 124
A Visual-Feedback Neural Subsystem 132
Summary of Biological Issues 136

8 Conclusions **139**
Scientific and Engineering Lessons Learned 139
Learning by Doing 139
Learning with Lookup-Tables 140
The Connectionist Architecture 142
Mental Models, Heuristic Search 143
Reflections on Brain and Behavior 143
Future Directions 144
Pragmatic Extensions 144
Theoretical Extensions 145

Bibliography **149**

Index **163**

Preface

We are fortunate to live at a time when several powerful intellectual disciplines are confronting issues whose understanding promises both fundamental insights into the nature of life and mind and the development of powerful new technologies. Phenomenal advances in computational power have allowed researchers to build systems whose complexity begins to approach that of interesting biological systems. As our ability to engineer such systems grows, the scope of possible models for biology, neurophysiology, and cognitive psychology is correspondingly broadened. A new kind of experimental paradigm is emerging in which ideas are tested by using them to engineer working systems. The success or failure of these constructions at performing the desired tasks provides a powerful means for eliminating insufficiently powerful concepts. As with traditional experiments, the scope of such systems is deliberately limited so as to bring the critical issues to the foreground. The present book describes a wonderful example of such an experiment.

The system, MURPHY, consists of a robot arm, a camera, and a controller which both looks at and maneuvers the arm. MURPHY's tasks are visually specified: "he" must grab targets while avoiding obstacles. His actions take place in the motor space of joint angles and he must somehow relate these to the visual space of his perceptions. As people do, MURPHY learns the relationship between what he does and what he sees by performing actions and building up a world-model from observations of their effects. MURPHY's world is a simple one in which he single-mindedly pursues his task, but in his combination of perception, action, and learning he combines ingredients that are essential to the functioning of all higher organisms.

It wouldn't be entirely surprising if progress in the understanding of these ingredients in some way recapitulates the steps that fledgling organisms must have taken as they pulled themselves out of the darkness. A major step on the road to consciousness is the development of internal representations to model the environment and to relate sensory and motor information. Learning allows such representations to be tuned both as the body of an individual organism develops and as evolution dramatically changes the basic somatic form and social environment. This robustness and flexibility is critical for systems that must deal with unpredictable environments. MURPHY's camera has auto-focus, causing the imaging geometry to change according to the location of the arm. Such a complexity is difficult to deal with by traditional means, but MURPHY cheerfully learns the relationship and successfully maneuvers in the face of this somewhat warped perception. This kind of flexibility and robustness is critical for new technologies whose environments are not predictable. Our ability to construct such flexible machines is critically dependent on our understanding of the basic issues discussed in this book.

MURPHY is one of the early organisms in the quest to repeat the biological journey in the engineering realm. He only begins to perceive the world dimly and act on it. But in these first steps we see the glimmer of things to come. As systems become more complex, we can expect both qualitatively new technologies and deeper insights into biological systems. Future pets, computer-human interfaces, movie actors, and recycling trash sorters may all be systems which build up their behavior by perceiving, acting, and learning. Future brain models may be so complex that they can only reside in computers and must be studied by techniques similar to those of neurophysiology. Some of these developments will yield deep insights, some may tremendously enhance human life, and others may cause social upheaval and pain to man's self-image. In any event, the journey will be stimulating and thought-provoking. The mix of biological modeling, psychological insight, hard engineering and computer science, and experimental exploration that is described in this volume may well be a model for future advances in this interdisciplinary endeavor.

Stephen M. Omohundro
International Computer Science Institute

Acknowledgements

Thanks are due to Subutai Ahmad, John Collins, Bill Greenough, Darrell Hougan, Bruce Lambert, and Steve Omohundro for their insightful discussions, to Andy Barto and Jim Houk for increasing my understanding of numerous issues in robot control and motor physiology, respectively, to Rob Shaw for his help in shooting "MURPHY: The Movie" (copies of 5-minute videotape available from author at cost of duplication), to Russ Shermer and Dave Ballman for cheerful systems support, to Norman Packard for encouragement and for his role as a provider and protector, and to committee members Narendra Ahuja, Jerry DeJong, and Larry Rendell (as well as Bill Greenough, Norm Packard, and Steve Omohundro). Extra thanks to Bill Greenough for contributing to numerous applications for funding and training during my period of thesis work, and for important background moral support throughout my graduate school years. Extra special thanks to Steve Omohundro for constant scientific and moral support in his capacity as thesis advisor, for his irrepressible enthusiasm, and for suggesting vision and robot kinematics as ideal domains for experimentation. This work was paid for in part by the National Center for Supercomputing Applications at Champaign, Illinois, National Science Foundation grant Phy 86-58062. Two additional sources of funding merit special mention. First, a Hewlett-Packard fellowship during four consecutive years of graduate school bought enormous time and freedom for scientific exploration, and it was during these years that the seeds for this book were planted. Second, warmest thanks are due to the University of Illinois Committee for Cognitive Science/AI fellowships, without whose support for two years completion of this work would have been difficult or impossible.

Regarding the making of this book, sincere thanks are due to Christof Koch for his encouragement and patience, and for providing the excellent working environment that allowed this book to be produced. Thanks also to Sari Kalin at Academic Press for her many helpful editorial suggestions, and for badgering me at just the right level to insure completion of the project. I am also indebted to Nan Boden for her careful proofreading of the manuscript. Lastly, thanks to Maria for her love and patience that have made the arduous task of bookwriting as continuously happy as possible.

Chapter 1

Introduction

It is unusual to find any single aspect of intelligent behavior that has elicited inquiry from so many different scientific perspectives, as has that of the visual control of arm movements. Researchers in robotics (see [159, 119]), AI [126, 29, 51, 93], connectionism [1, 135, 143, 47, 79, 172, 115, 180, 98, 199], motor physiology [86, 26], motor psychophysics [11, 83], motor learning (see [186, 195]), perceptual [21, 73] and cognitive [160, 204, 34, 82, 125, 211, 21] development, neuropsychology ([176], see [66]), and neurophysiology [151, 92, 153, 61, 62] have marshalled the tools of their various trades in an attempt to better understand aspects of this "simple" sensory-motor process. Vision-based control of a multi-link arm is in fact a very difficult computational problem, requiring an understanding of the often complex spatial relationships between the body and its parts and the coordinate systems of the external world. As is true for the visual control of eye-movements or locomotion, the ability to reach for visual targets requires that parts of the body be "lined up" through sequences of *motor* commands with an external environment that is specified only *sensorially*. A profound and very general problem thus exists as to how to intelligently relate information in these two different internal languages. The need to *learn* these relationships may also be of great benefit, since the needed sensory-motor mappings for visual control of an arm are complex, system specific, and can change over time in unexpected ways.

Figure 1.1: MURPHY's workspace at the Center for Complex Systems Research at the University of Illinois. The robot arm is a 5 degree-of-freedom Rhino XR-3 robot arm moving in the image plane of a JVC color video camera. Simple image processing was performed by a Datacube board set. The Sun workstation pictured was used to control all aspects of MURPHY's behavior.

A robot-camera system called MURPHY[1] has been developed to explore a new approach to the kinematics of sensory-motor learning and control for a multi-link arm. The problem of central concern in this book is that of guiding a multi-link arm to visual targets in a cluttered workspace. In contrast to other formulations of the motion-planning problem [16, 29, 30, 51, 93, 104, 126], emphasis in this work has been on "ecological" solutions that are more relevant to the typical visually-guided reaching behaviors of humans and animals in natural environments, than to the highly structured demands of an industrial setting. For example, we seek algorithms that work well in unmodeled workspaces whose effective layouts can change from moment to moment with movements of the eyes, head, limbs, and body. Further, we are primarily interested in computing *good* rather than optimal paths for the arm, since relaxation of optimality as a performance criterion makes possible a much simpler class of motion-planning algorithms. Exact kinematic tolerances are also considered less important than the flexibility associated with the ability to learn kinematics from sensory-motor "practice".

While not intended as a strict model for a particular neural subsystem in a particular animal species, MURPHY's data structures and algorithms have been constrained to adhere to the *style* of representation and computation seen in biological sensory and motor systems to the greatest extent possible. Three guidelines have been used to constrain system design. First, we assume no *a priori* model of arm kinematics, or of the imaging characteristics of the camera/vision system. Instead, MURPHY must learn "by doing" at the start, and remain plastic to subsequent changes in his visual-kinematic relations. Second, we exclude sophisticated built-in algorithms or special-purpose data structures for obstacle avoidance, such as the ilk of powerful geometric techniques that have been proposed by others (e.g., [16, 29, 30, 51, 93, 104, 126]). Rather, we seek out a virtual ma-

[1]The name "MURPHY" was taken from the movie *Robocop* (Summer 1987), and was suggested to the author by David Ballman. It is thoroughly overused throughout this book. The author agonized over the choice of pronoun for MURPHY, whether "he", "it", or worst of all, "the system". It was decided on purely linguistic grounds the name MURPHY is more of a "he" than an "it"; since personal pronouns are also used to refer to pets, ships, and robots in the movies, it was felt this was not a case of unprecedented or reckless anthropomorphization.

Figure 1.2: Illustration of a reaching problem with obstacles. At the top is a reproduction of the original slide used to propose the connectionist planning algorithm to the author's thesis committee. Below is the solution path from one of MURPHY's first successful reaching episodes.

chine architecture within which simple parallel and sequential operations relevant to vision-based motion planning are readily realizable. Third, we consider a much richer and more direct role for visual processing than is usually assumed in robot motion-planning systems. For example, we assume no foreknowledge (e.g., symbolic description) of the locations and shapes of workspace obstacles, and avoid non-ecological representational biases for specific classes of visual objects (e.g., polygons). The purpose of these three constraints is to force our design into a niche in the space of algorithms that is more akin to the solutions at work in the brains of behaving animals.

The resulting connectionist system represents a significant departure from analytic-geometric approaches to robot control and motion planning. First, both forward and inverse-differential kinematics are learned from scratch during an initial training period, in which the arm is "flailed" in front of the camera, and during subsequent periods of purely "mental" practice. Then, MURPHY uses best-first heuristic search with mental images of his arm in the workspace to find paths for his arm amidst obstacles. The present work differs most from other approaches to robot motion planning in its representational and algorithmic simplicity. For example, both MURPHY's joint space and visual workspace representations consist of topographically-mapped populations of simple neuron-like units, and all needed kinematic relations are represented as a single layer of modifiable synaptic connections between appropriate combinations of these populations. This architecture allows MURPHY to directly generate and manipulate an explicit full-visual-field image of his arm in the workspace, upon which visual operations relevant to motion planning, such as collision detection, become single, parallel virtual-machine operations. No other intermediate geometric data structures are used in this approach. Also, no inverse kinematic model of the arm is used at any stage of planning or control. Several recent connectionist systems have been described that learn arm kinematics and dynamics [1, 47, 98, 199, 115, 143, 180], two of which share representational features in common with MURPHY [47, 115], but this author knows of no other connectionist approach to motion planning for a multi-link arm. MURPHY's basic system organization and behavior have been described in two earlier reports [135, 136].

Chapters 2, 3, and 4 contain descriptions of MURPHY's physical setup, connectionist architecture, learning algorithms, and performance on a variety of visually-specified reaching problems. In chapters 6 and 7, we explore the degree to which MURPHY's representations and algorithms are relevant to the psychology, physiology, and neural bases for the control of directed, voluntary arm movements in humans and animals. A number of intriguing correspondences are described, including several that were unanticipated by the author at the time of MURPHY's design.

Chapter 2

MURPHY's Organization

The Physical Setup

MURPHY's algorithms and data structures constitute a new approach to vision-based robot arm control. In order to ensure both representational and task realism to the greatest degree possible, MURPHY has been implemented with a real camera and a real robot arm. MURPHY's physical setup consists of a JVC color video camera pointed at a Rhino XR-3 robotic arm (fig. 2.1). Only the shoulder, elbow, and wrist joints are used, such that the arm can move only in the 2-D image plane of the camera, at a distance of 7 feet. (A fourth, waist joint is not used). White spots are attached to the arm in convenient places; when the image is thresholded, only the white spots appear in the image. This arrangement allows continuous control over the complexity of the visual image of the arm, which in turn affects computation time during learning (for a serial computer). A Datacube image processing system is used for the thresholding operation and to "blur" the image in real time with a gaussian mask. The degree of blur is variable and can be used to control the degree of coarse-coding (i.e., receptive field overlap) in the camera-topic array of visual units, as described below. Joint angles are software controllable, with a stepper motor for each joint. Figure 2.1A shows MURPHY's view of his arm, a target (white cross stuck to backdrop), and obstacles (other white paper cutouts). Figure 2.1B shows MURPHY after successfully reaching for the target, i.e., with hand visually

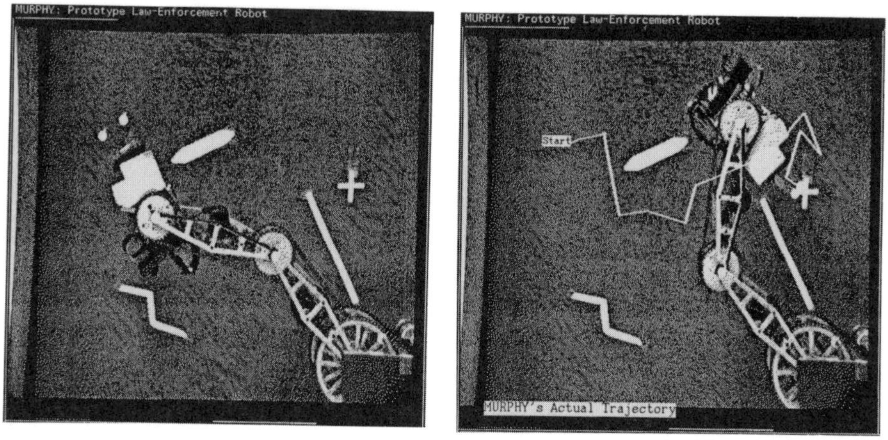

Figure 2.1: The physical setup. A. MURPHY's view of his arm in a typical starting configuration. Target (white cross) and obstacles (other white shapes) are paper cutouts stuck to backdrop 6" behind the plane of the arm's movement. B. MURPHY after having successfully reached for the target, i.e., with hand in visual superposition with the target. The path connecting the start and goal configurations is collision free if no part of the arm falls into superposition with any obstacle at any point along the trajectory.

superimposed with the target from the perspective of the camera. A collision is said to occur when any part of the arm falls into superposition with any of the backdrop obstacles.

The Connectionist Architecture

It has been noted that connectionism is centrally concerned with capturing the *style* of computation seen in nervous system structures, in contrast to the more conventional models of computation that underlie the modern digital computer [187]. As such, MURPHY's systems-level architecture has been based on the observation that a large fraction of even the most highly evolved brains is devoted to the explicit representation of the animal's sensory and motor state, exemplified by the several dozen sensory, higher sensory, and motor areas of the primate cerebral cortex [28, 140, 68]. During normal behavior, each of these neural representations carries behaviorally-relevant state information, giving rise to a rich set of online associative learning opportunities among themselves. These distinct unit populations have the possibility of interacting bidirectionally through long-range excitatory interconnection pathways in the brain; following a basic tenet of connectionist computation, it is likely that these synaptic pathways are an important locus for associative learning.

MURPHY's connectionist architecture is similarly based on several interconnected populations of neuron-like units (fig. 2.2), encoding positions and velocities in both the visual and joint domains:

1. *Visual-Field Population.* The principal sensory population is organized as a rectangular, visuotopically-mapped 64 x 64 grid of visual units, each of which responds when a visual feature (such as a white spot on the arm) falls into its receptive field (fig 2.2 upper left, fig. 2.3A).

2. *Hand-Velocity Population.* The second visually-activated population of 24 units encodes the direction and amplitude of MURPHY's hand motion through the visual field (fig. 2.2 lower left, fig. 2.3B). These units are thus "fired" only by the distinct visual image of the hand, and are selective for the particular

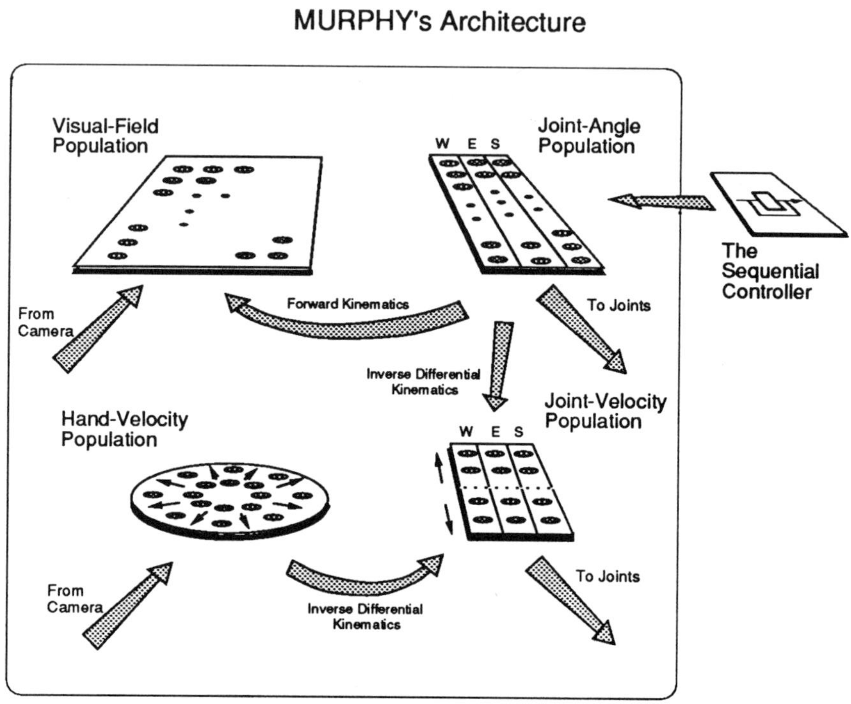

Figure 2.2: MURPHY's machine architecture. Four interconnected populations of neuron-like units implement MURPHY's forward-kinematic and inverse-differential kinematic mappings. Visual input from camera enters at left; motor signals to joints exit at right. A sequential controller (upper right) drives the connectionist joint-angle population during both learning and performance phases.

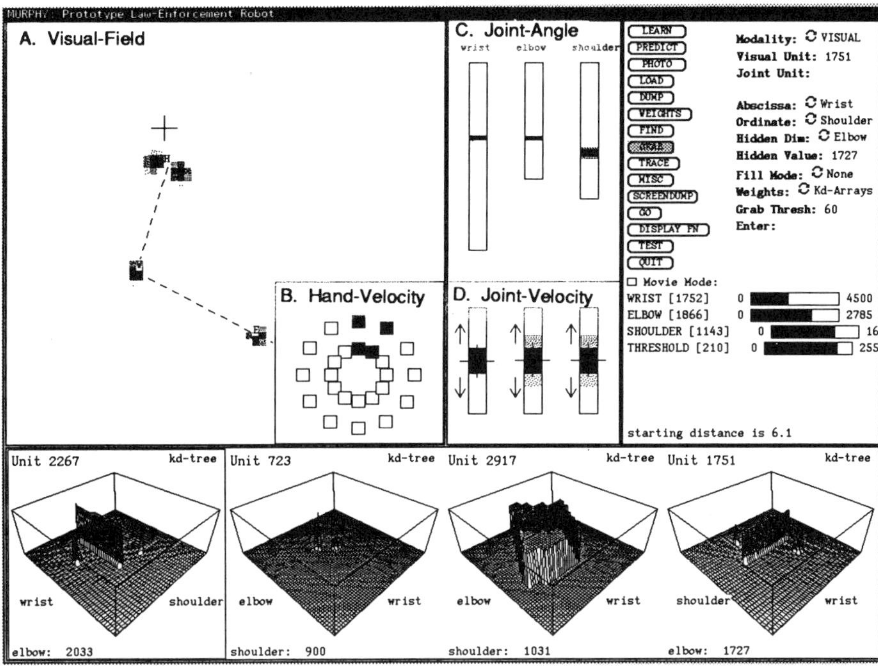

Figure 2.3: MURPHY's population activity during simulation. A. Looking down on the 64 x 64 cameratopically-organized map of visual units, with grey level proportional to unit activity in response to a view of the arm. The dashed lines connecting the visual blobs on each joint represent the link volumes for purposes of collision avoidance, and are explicitly computed as a post-processing step. B. Direction-selective visual population responds to the motion of the hand through the visual field. Also used to encode the *desired* motion of the hand as MURPHY reaches for a visual target. C. Three joint-angle subpopulations encoding the positions of the wrist (left), elbow (middle), and shoulder (right). D. Three joint-*velocity* subpopulations encoding direction and magnitude of joint perturbations during arm movements. Graphs at bottom show 4 visual-field units' activation functions plotted over 2 of 3 joint variables.

direction and amplitude of hand motion through the visual field as MURPHY moves his arm in the workspace. The hand is detected as the largest contiguous blob of activity on the retina; however any other distinctive cue could be employed.

3. *Joint-Angle Population.* The principal joint population of 273 units consists of three subpopulations encoding static joint angles; the angle of each joint is coded individually in a subpopulation dedicated to that joint (fig. 2.2 upper right, fig. 2.3C).

4. *Joint-Velocity Population.* The second joint population of 24 units also consists of three subpopulations, each coding the *velocity* (i.e., change in angle during a unit timestep) for one of the three joints (fig. 2.2 lower right, fig. 2.3D).

Representational Choices

The way in which bits of "meaning" are assigned to individual units is possibly the single most important design decision in a connectionist learning system, having a profound impact on virtually all aspects of system learning and performance behavior. The representational choices detailed above are therefore not accidental, but carefully designed to facilitate MURPHY's visual-motor learning tasks.

Coarse-Coding

All four of MURPHY's populations are coarse-coded, i.e., consist of units with peaked *response fields* (i.e., receptive and/or projective fields) that overlap among neighboring units, and whose peak sensitivities vary systematically from unit to unit across the population (fig. 2.4). For example, each visual-field unit has a gaussian sensitivity profile over a localized region of the image that overlaps with several of its neighbors. Likewise, each unit in the joint-angle population has a peak response for some joint angle, and neighboring joint units within a joint subpopulation have overlapping "projective fields" and progressively increasing joint-angle centers. The gaussian receptive field profiles of the visual field units were easily derived

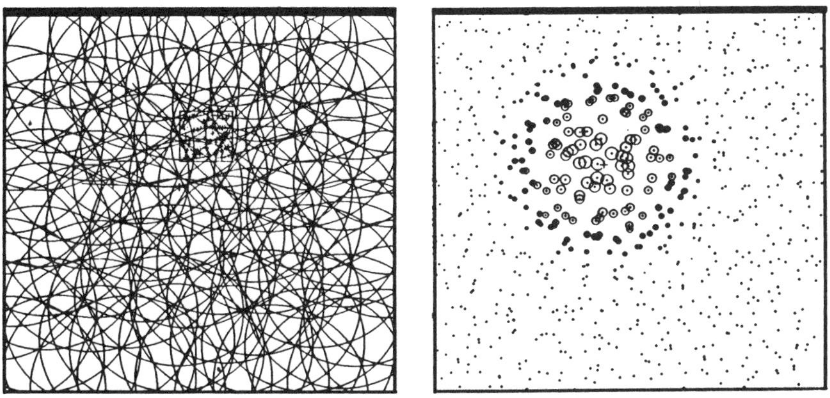

Figure 2.4: Coarse-coding makes efficient use of noisy units for the representation of multidimensional spaces. At left, a two-dimensional input space is covered by an array of overlapping binary receptive fields. At right, a "cloud" of activity across this coarsely-tuned array is shown for a single input stimulus, where circle size corresponds to degree of unit activation. Contrary to intuition, as receptive fields grow larger, the ability to localize an input can actually *increase* (see [80, 48]). This effect is essentially due to the increased number of receptive fields activated by a single input, giving rise to more "votes" as to the input's location [130]. However, as receptive fields grow *too* large in an input space, they begin to act as low-pass filters that can strongly limit the derivatives in the responses of neurons in subsequent representations [137].

from the hardware image convolution operation provided by the Datacube boards. For reasons of efficiency, receptive field profiles in the other three populations were chosen to fall off *linearly* with distance from their centers, thresholded at zero. Reasonable choices for receptive field size, density, and profile for each of the populations almost always lead to reasonable system performance, though a rather complex set of engineering tradeoffs can make the "optimization" of these parameters nontrivial and highly system dependent.

This general type of coarsely-tuned representation is ubiquitous in biological sensory systems and has been attributed a number of representational advantages, including finer stimulus discrimination and increased immunity to noise and unit malfunction [15, 80, 48, 12, 187, 137]. Coarse-coding has also been shown to contribute to continuity-based generalization in connectionist learning systems [14, 209, 149, 135, 137]. These issues are taken up in greater detail in chapter 3 in the context of the synaptic learning rule.

Level of Decoding

A related representational issue with important consequences for learning concerns the degree to which the input space is *decoded* in the activities of individual units. Definitionally, connectionist networks assume that the transformation of patterns from input to output in an associative learning problem are captured entirely by a set of positive or negative synaptic weights that interconnect the simple processing units. It may therefore be observed that the ease of learning a particular function in a connectionist network will depend on the degree to which the bit of meaning coded by each output unit is related in a simple fashion to the bit of meaning coded by each input unit upon which it depends. At one extreme, the input representation might be completely *decoded*, consisting of a separate unit for each possible input pattern. While this scheme requires a combinatorially explosive number of units, only one of which is ever active at a time, and the system must "see" every possible input pattern during training, the actual weight modification rule is rendered very simple. This is the case since each input unit stands for a single, specific input pattern; for each output unit that "likes" that input pattern, i.e., that is intended to fire strongly in the presence of that input pattern, the input unit can simply drive that output unit directly with an excitatory synapse (i.e., positive weight). Conversely, output units that "dislike" the given input pattern can be driven with an inhibitory synapse (i.e., negative weight). At another extreme, the single unit representation can be chosen in a highly *encoded* fashion, where the activation levels of the output units depend on complex, nonlinear interactions among input units. In this case, since the relation between an input and output unit cannot be captured as a simple positive or negative scalar, it is necessary to use a more powerful network representation, consisting either of multiple layers of weights and processors, or a single layer of processors capable of computing more complex (than linear) functions of their inputs. Both types of more powerful network learning schemes have been proposed, giving rise to such well-known algorithms as back-propagation [182, 120, 157], the Boltzmann machine [81], and a variety of higher-order learning networks [17, 63].

An intermediate coding strategy is exemplified by MURPHY's joint-angle population: the values of the underlying input parameters are decoded one dimension at a time (i.e., by joint—see figs. 2.2, 2.3). This type of partial decoding avoids the combinatorial explosion seen in the case of complete decoding, which entails a separate unit for every possible global joint configuration (or a representative subsample). This number grows exponentially in the number of joints. Instead, a global joint configuration can be captured as a conjunct of *three* joint units (or in general d units for a d-dimensional space of inputs), one from each of the joint subpopulations. If, for example, a functional mapping were to be learned from the joint-angle population (acting as input) to the visual-field unit population (acting as output), it is clear that the effect of a given joint unit on a given visual-field unit is clearly not as simple as a scalar, since the visual-field effect of one joint angle depends on the angles of the other two joints. One way of capturing this nonlinear conjunctive interaction among input units would be to insert one or more intermediate layers of nonlinear units (such as the thresholded linear unit, or TLU) between the joint and visual unit populations, thereby requiring one of the powerful, multi-layered synaptic learning rules that have been developed in recent years (e.g., back-propagation [182]). The approach taken in this work, however, has been to use another type of simple connectionist unit, the *sigma-pi* unit, that computes multiplicative conjunctions as an elemental operation (see chapter 3). To a first approximation, *sigma-pi* units are a means of replacing a combinatorial explosion of input units with a combinatorial explosion of output synapses. Where synapses are smaller, "metabolically" cheaper, easier to create on the fly than full-blown units, and available in prodigious quantities, the tradeoff is a good one. Using *sigma-pi* units is thus an effective scheme for "faking" a completely decoded input representation in return for the drastically simpler synaptic learning rule associated with this type of decoded representation.

This use of conjunctive (i.e., multiplicative) terms at the input to a neuron is essentially a means for constructing localized receptive fields of higher dimension from several receptive fields of lower dimension. In the example mentioned above, each *sigma-pi* conjunct corresponds to a localized 3-dimensional receptive field in joint space, built from three 1-dimensional joint-space receptive fields. This use of locally-tuned receptive fields brings the current approach in direct correspondence with methods of Radial Basis Functions [31, 148, 149, 163], discussed in chapter 3, and is essentially a mechanism for constructing a locally-smoothing neural lookup table.

MURPHY's Kinematics

Traditional Methods

Before delving into MURPHY's learning procedure in detail, we first consider the abstract functional roles played by each of MURPHY's populations, in relation to the traditional language of robot *kinematics.* A classic task in robotics is to send the "hand" on a multi-jointed arm to a visual target, ignoring for the moment the possibility of obstacles in the workspace. The traditional approach involves first writing explicit kinematic equations for the arm based on its specific geometric details:

$$K : \Theta \mapsto X_{\mathbf{ws}},$$

where K maps joint angles Θ into the workspace coordinates $X_{\mathbf{ws}}$ of one or more control points on the arm. When a camera is inserted in the system, additional equations are needed to model the transformations of workspace coordinates due to the imaging characteristics of the camera. In general, however, it is most often useful to specify where the hand *should* be, and to compute the joint angles necessary to achieve this goal. This involves the solution of the kinematic equations to generate an *inverse*-kinematic model:

$$K^{-1} : X_{\mathbf{ws}} \mapsto \Theta.$$

The algebraic derivation of the inverse-kinematic equations has been called "the most difficult problem we will encounter" in vision-based robotics [159], so much so that robot arms are specifically designed to make these calculations as simple as possible. The major difficulty with inverse kinematics stems from the fact that problem is ill-posed. In other words, there are in general many (possibly infinite) solutions (i.e., settings of the joint angles Θ) that will put the gripper of a redundant arm at a given point in the workspace. A third useful relation is the inverse *differential* map,

$$\mathcal{J}^{-1} : \Theta, dX_{\mathbf{ws}} \mapsto d\Theta,$$

taking joint angles Θ and a desired workspace perturbation $dX_{\mathbf{ws}}$ for one or more of the arm's control points into a set of joint-angle perturbations $d\Theta$. Unlike the general inverse-kinematic relation K^{-1}, this inverse map is linear since it is concerned with infinitesimal perturbations.

When the forward-kinematic model is accurate, and the inverse models can be algebraically derived, the actual computations necessary for arm control are relatively fast and easy to carry out on a digital computer. Serious difficulties arise, however, when either i) the arm or the visual imaging system does not perfectly conform to the simplifying geometric and trigonometric assumptions embodied in the kinematic equations, ii) the structure or material composition of the arm makes its explicit kinematic analysis difficult or impossible (e.g., a flexible arm), or iii) the kinematics change over time, such as through growth, aging, wear, or structural damage. In these cases, the arm can crash. When obstacles are present in the workspace that further constrain the possible configurations of the arm, altogether different algorithmic techniques are required for planning arm trajectories [16, 126, 29, 154, 93, 119]. To illustrate the practical difficulty of achieving an accurate analytic model of even a simple arm-camera system, we note an interesting and rather quirky complexity that is present in MURPHY's physical setup. MURPHY's camera has autofocus; when any part of the arm happens to fall in the focusing field of the camera, the camera shifts focus to the depth of the arm; otherwise it remains focused on the background. This erratic shift in focus causes a significant optical expansion and contraction that alters the apparent locations of objects in the peripheral visual field.

Based on the difficulties of the explicit analytical approach, it is highly desirable for a mobile agent to be able to *learn* a model of its sensory-motor environment from scratch, in a way that depends little or not at all on the specific parameters of the motor apparatus, the sensory apparatus, or their mutual interactions. As an alternative to inverting kinematic equations analytically, therefore, a number of authors have proposed connectionist systems that learn inverse kinematics by interacting with the environment and observing the relationship between hand coordinates and the associated joint angles [1, 115, 199]. However, a system that must learn to produce a *single* solution must learn therefore to reject all other possible solutions. One approach is to train the arm in only a single joint configuration per final position of the hand, effectively discarding the arm's excess degrees of freedom. This solution is undesirable when obstacles are present in the workspace, since the excess degrees of freedom may be necessary for the avoidance of obstacles. Furthermore, in this case it is no longer possible to simply solve for the final configuration of the arm in reaching for a target, since the path of the arm from start to finish becomes critical. In a slightly different approach, Rumelhart proposed the idea of learning an inverse model by back-propagating error signals through a fixed network previously trained with a *forward*-kinematic model [180], an idea developed further by Jordan [98]. This approach allows constraints, such as smoothness of trajectory through a sequence of arm states, to "take up" the extra degrees of freedom in the inverse solution space [98]. By analogy with the analytic, equation-based approaches, most connectionist models applied to kinematic learning tasks to date have mapped problem-space variables such as workspace positions and joint angles directly onto connectionist units [1, 47, 199, 98]. In contrast, Kuperstein's system learns inverse kinematics using highly decoded representations for all system variables [115].

Connectionist Kinematics

The Forward Model

MURPHY conceptualizes the problem of reaching for a visual target in such as way as to avoid the construction of an inverse-kinematic model (although the much simpler inverse-*differential* map is used below). The underlying principle is that a solution to a difficult inverse problem can often be achieved with simple heuristic search on a *forward* model. Thus, MURPHY's four populations implement two basic visual-kinematic relations[1]. The primary relation is encoded in the map from the joint-angle population to the visual-field population, giving rise to a connectionist version of the traditional forward-kinematic map:

$$K : \Theta \mapsto X_{\text{vf}},$$

where joint angles Θ are mapped into full-visual-field images of the arm X_{vf} rather than into explicit coordinates for a set of control points. This map is encoded by the set of modifiable synaptic connections from the joint-angle population to the visual-field population. Running the forward model involves first setting up a state of activity on the joint-angle population that encodes some global arm configuration, and then allowing the synaptic projection to the visual population to generate a visual image of the arm in that configuration (see fig. 3.1). For visually-specified problems, such as sending the hand to a visual target in the presence of obstacles, an *analogical*

---

[1]The term *visual-kinematic* is used here to denote the composition (for both forward and inverse-differential maps) of the trigonometric kinematic functions, relating joint angles to workspace coordinates, with the visual functions relating workspace coordinates through the camera and vision hardware to the coordinates of MURPHY's internal visual representation.

visual representation is a very convenient data structure, allowing a range of visual heuristics to be efficiently computed. It was hypothesized that the ability to generate full-blown "mental" images of his arm in arbitrary configurations would allow MURPHY to internally "flail" his way to a target while avoiding obstacles, using all of the arm's available degrees of freedom. The use of this type of analogical visual representation of the arm and the workspace as the primary substrate for motion planning differentiates MURPHY from the usual methods of robot kinematics and path planning. This issue is taken up at length in chapter 5.

As shown in figure 3.1, the thresholding operation applied to the video image results in a significantly reduced visual image of the arm, consisting of only several disconnected visual blobs corresponding to the white spots place on the gripper and on each of the joints. MURPHY can, alternatively, be provided with the *complete* image of the arm during training, including links, such that his forward-kinematic model generates full visual mental images of the arm instead of the reduced images of fig. 3.1. However, the space-time requirements of the current non-parallel implementation grow linearly in the area of visual activity, making the explicit postcomputation of link volumes a more attractive solution. (A fully parallel implementation consisting of one processor per visual-field pixel would run in constant time as a function of visual field load, eliminating the slowdown, though the memory requirements would still increase.) During planning, where a complete image of the arm is needed for collision avoidance, the visual blobs on each of the joints are interconnected with straight lines, crudely representing the volumes of the links (fig. 2.3A).

Inverse-Differential Kinematics

In addition to learning forward kinematics, MURPHY is also able to represent the inverse *differential* kinematic map,

$$\mathcal{J}^{-1} : \Theta, dX_{\mathrm{hand}} \mapsto d\Theta,$$

which takes the current joint configuration Θ and a desired differential movement of the hand dX_{hand} through the visual field into an appropriate set of joint perturbations $d\Theta$. This map is encoded by a combined projection from the joint-angle population encoding Θ and the hand-velocity population encoding dX_{hand} to the joint-velocity population which encodes small angular perturbations $d\Theta$ to the joints (see fig. 2.2). It is run in the same way as the forward-kinematic map described above, i.e., by setting up initial states on the two input populations and allowing the synaptic projections to generate an "image" on the output population (in this case a joint-angle perturbation vector). This procedure allows MURPHY to steer his hand in a desired direction through the visual field. Elsley [47] has also described a neural network that learns the inverse-differential map for a simulated 2-degree-of-freedom (*dof*) arm, whose accuracy exceeds that of an analytically-derived controller.

As for the case of the standard inverse-kinematic relation, $\mathcal{J}^{-1}$ is not unique, since there are in general an infinite set of perturbations to the three joints that will result in a given direction of hand motion through the visual field. This problem of multiple inverse solutions occurs whenever a forward map f takes R^n to R^m where $m < n$, i.e., collapses a higher dimensional space onto a lower dimensional space. In MURPHY's case, for a given global joint configuration, the forward-differential kinematic map $\mathcal{J}$ takes R^3 (i.e., 3 joint velocities) into R^2 (i.e., the 2-d space of directions in the visual field). The inverse map $\mathcal{J}^{-1}$ for a given joint configuration therefore takes R^2 back up to R^3; the inverse image of a given direction of hand motion therefore consists of an infinite set of possible joint perturbations. Which single solution will be learned and produced? Differential kinematics is linear (by definition), such that if a joint perturbation $d\Theta$ gives rise to a hand movement of dX then a joint perturbation of $\alpha \cdot d\Theta$ will give rise to a hand movement of $\alpha \cdot dX$. Consequently, $\mathcal{J}^{-1}$, while not unique, has the attractive property that the *average* of the infinite set of solutions to a given inverse problem is also a solution[2]. Thus, the pattern of activation induced on MURPHY's joint-velocity population

[2] A linear map can rotate, scale, and sheer the domain space, and/or collapse the domain along any linear subspace. When there is a reduction in dimension from domain to range, each point in the range has as its inverse image a linear subspace in the domain, such as a line or a plane. One property of a linear subspace is that the average (i.e., center of mass) of the subspace also lies in the subspace.

will represent a superposition of those solutions seen during training, and its "center of gravity", or average, can itself be interpreted as an approximative solution. Further, since the resulting solution is the average of those in the inverse image, which is a bounded linear subspace, it can at worst be the solution of average magnitude and at best the solution of *least* vector magnitude among all possible solutions, i.e., the least perturbation to the joints that gives rise to a unit-length move of the hand in the desired direction. This is a consequence of the following assumptions: i) the domain from which training examples are uniformly drawn is a convex, bounded volume centered on the origin (e.g., the unit sphere in MURPHY's 3-d space of joint velocities), ii) the forward-differential kinematic mapping is linear (by definition), such that the inverse image of a range point is a bounded linear subspace in the domain (e.g., a line embedded in this 3-d unit sphere), and iii) the inverse solution produced is the average, or center of mass, of the points lying in this linear subspace, and can have at most average vector length.

As a final note on the subject of kinematics, it may be asked why a standard inverse-kinematic model of the arm-camera system could not be implemented in MURPHY's populations. MURPHY *can* in fact learn the inverse map from the visual image of his arm X_{vf} into its corresponding joint-angle image Θ. Unfortunately, for the purposes of acquiring a visual target, this particular map is not as useful as it might be: MURPHY has no means of generating *in advance* a full visual image of his arm grasping a visual target, the needed input to this inverse map.

The Sequential Controller

During both his learning and performance phases to be described in subsequent sections, MURPHY is able to carry out simple sequential operations driven by an unelaborated controller external to his connectionist architecture (fig. 2.2). Since these sequential control functions and their "neural" implementations currently fall outside the scope of this project, they have been compartmentalized and kept as simple as possible in order that the bulk of MURPHY's competence derive from portions of the model that *have* been fully elaborated.

Chapter 3

How MURPHY Learns

MURPHY's learning behavior may be broken into two independent levels of description. The first concerns the teacherless associative-learning paradigm called *learning by doing*, which deals at a macroscopic level with the ways in which MURPHY extracts information from his environment. The second, microscopic, level of analysis concerns the way in which MURPHY's kinematic relationships are actually learned and encoded at the level of the synapse.

Learning by Doing

Perhaps the most elusive goal in machine learning over the past decade has been to eliminate the need for the participation of an intelligent teacher or programmer in the learning process. Thus, learning has often been accomplished by presenting either positive or negative examples to the learning system in the form of input-output pairs, or, alternatively, has relied crucially upon knowledge structures built in by the programmer. Where learning has been *unsupervised*, as exemplified by clustering algorithms, it has generally been limited to detecting *a priori* statistical structure in a stream of input patterns (e.g., [110, 184]).

It has seemed an inevitable tradeoff that a system needing to rapidly learn specific, behaviorally-relevant input-output mappings must necessarily do so under the auspices of an intelligent teacher with a ready supply of task-relevant training examples. This state of affairs has seemed somewhat paradoxical, since the perceptual and cognitive development of human infants does not depend on the moment by moment intervention of a teacher of any sort.

A major goal of this work has been to explore the limits of a learning procedure called *learning by doing*, that has thus far been relatively little exploited in the field of machine learning. Learning by doing is particularly well suited to acquiring knowledge in reactive environments, and is described here in the context of sensory-motor behavior. In its basic form, the learning agent

i) begins with a repertoire of actions and some form of sensory input,

ii) exercises its repertoire of actions, thereby building a mental model that relates actions to their sensory consequences and vice versa, and

iii) runs its mental model, in one or both directions, to "envision" solutions to problems posed by the environment.

To consider a concrete example, MURPHY builds his forward kinematic model in exactly this way. By stepping his arm through a small, uniform sample (approximately 17,000) of the 3.3 billion legal arm configurations, MURPHY learns the relationship between his joint angles and the state of his visual field, by modifying weights between the joint-angle and visual-field populations. During training, the particular sensory and effector hardware to which MURPHY is physically wired (i.e., the video camera and the robot arm) *enforces* some mapping between the states of these two populations. The mapping comprises both the kinematics of the arm and the optical parameters and global geometry of the camera/imaging system, including the erratic autofocus phenomenon described in chapter 2. After training, a state of activity on the joint-angle population generates a "mental image" on the visual-field population via the weighted interconnections developed during training. The mental image is a picture of the arm in the given joint configuration, approximating that which would have been produced directly by the camera (fig. 3.1). In his current state, MURPHY steps through his entire joint

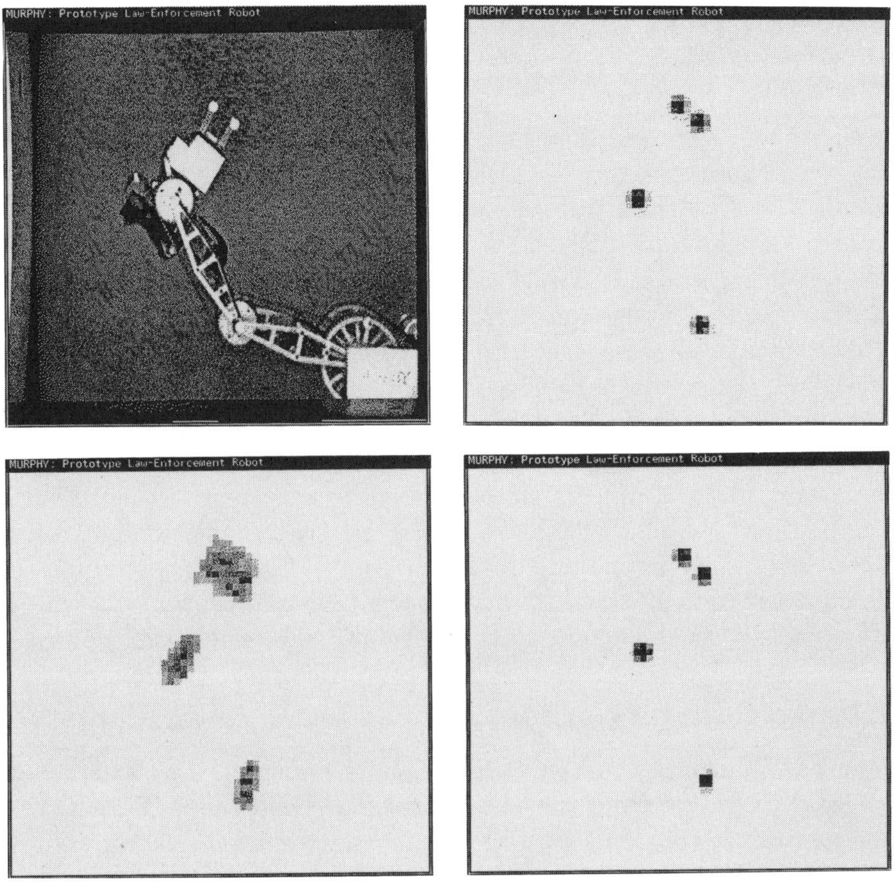

Figure 3.1: Four visual representations. The upper left frame shows the unprocessed camera view of MURPHY's arm. White spots have been stuck to the arm at various places, such that a thresholded image contains only the spots. This allows continuous control over the visual complexity of the image. The upper right frame shows the resulting pattern of activation over the 64 x 64 grid of coarsely-tuned visual units as driven by the camera input pathway. The lower left frame depicts an internally-produced "mental image" of the arm in the same configuration, as driven by weighted connections from the joint-angle population. Note that the mental trace is a sloppy but recognizable approximation to the camera-driven trace. The fourth frame shows the mental image generated using k-d trees in the place of *sigma-pi* units (see pp. 48-50). The lower two frames thus exhibit the output of the forward-kinematic map under two representational conditions.

space in around 5 hours, developing a total of approximately 2.5 million weights from the joint to visual-field populations.

In the above example, each visual-field unit is fired *unconditionally* by its input line from the camera during learning, while the weights on the "conditioned" input lines from the joint-angle population are appropriately modified[1]. It is thus useful for terminological reasons to note that, from the perspective of an output unit (in this case a visual-field unit), this type of learning is *fully supervised*, in that each output unit is told exactly when and how strongly to fire during learning by its unconditioned input line acting as a teacher. However, in comparison to the standard use of supervised learning methods, the crucial property of learning by doing is that no outside *intelligent* teacher is used to provide input-output pairs for learning or to build in specific prior knowledge as to the nature of the environment. Laws of physics simply translate actions into their resulting percepts, both of which are represented internally. The learning agent need only notice and record these relationships for later use. In contrast to traditional *unsupervised* learning approaches, learning by doing allows the acquisition of specific, task-relevant mappings.

Abridged History of the Idea

The psychological concept of learning by doing may be attributed at least to Piaget from the 1940's (see [65] for review). Piaget, the founder of the *constructivist* school of cognitive development, argued that knowledge is not given to a child as a passive observer, but is rather discovered and constructed by the child, through active manipulation of the environment (from Ginsburg & Opper [65]):

> "One of Piaget's central themes is that concrete action precedes and makes possible the use of intellect... As the infant begins to manipulate the objects which surround him, he gradually develops a practical "understanding" of external reality... Thus,

[1]The terms "conditioned" and "unconditioned" are drawn from the tradition of *classical conditioning*. An unconditioned input is one that results in an innate response, such as the camera input line to a visual-field unit, whereas a conditioned input becomes an effective stimulus only as a result of pairing with the unconditioned stimulus. In the current context, this corresponds to the input from the joint-angle population to a visual-field unit.

> the acquisitions of the sensorimotor period form the foundations of the individual's mental development..."

A handful of workers in artificial intelligence have addressed the issue of learning by doing, though only in highly schematized, simulated domains, where actions and sensory states are represented as logical predicates [22, 173, 37]. Barto & Sutton [19] demonstrated how a simple simulated automaton with two actions and three sensory states can build a model of its environment through exploration, and subsequently use it to choose among behavioral alternatives. In previous work, this author [134] described a connectionist model that learned to mentally rotate 3-d simulated wire-frame objects by first actively manipulating a series of practice objects. Rumelhart [180] and Jordan [98] have suggested in the context of robot kinematics that a feedforward model can be learned through practice; once learned, the forward-going "mental model" can then be used internally to train an *inverse* model using back-propagation. Several other authors have also recently described connectionist systems that learn robot arm kinematics [135, 1, 47, 199, 115]. Jordan & Rosenbaum [99] review a variety of similar approaches to sensory-motor learning.

Learning by doing is akin to the very old concept of *system identification* from control theory, where an unknown plant is identified from examples of its input-output behavior. These mathematical parameter-estimation techniques have typically assumed relatively rich *a priori* knowledge as to the structure of the underlying process to be identified (e.g., linearity), thereby fixing a relatively small space of parameters in which to search. Atkeson [9] has used such a technique to estimate inertial and load parameters for a multi-link robot arm. Raibert [166] demonstrated that by sending torque commands to the joints of a robot arm and observing the resulting joint accelerations, it was possible to fill in a *table* of inertial and other force parameters that could be used to control a robot arm. While this approach still relied in part on knowledge of the equations of motion governing arm dynamics, it demonstrated the utility of tabular approaches to system identification that may rely little or not at all on advance knowledge of the form of the unknown functions. Albus [3] proposed the CMAC model as a pure tabular approach to learning by doing in sensory-motor control, and Miller et al. [143] have implemented this idea to learn high-performance control of the dynamics of a 5-*dof* robot arm.

In summary, each of the research traditions of psychology, artificial intelligence, connectionism, and control theory have recognized in its various incarnations the utility of learning by doing for the extraction of information from unknown reactive environments.

Motivating Sigma-Pi Learning

To this point, we have seen that MURPHY's architecture, consisting of four interconnected populations of visual and motor units, is specifically designed to take advantage of associative learning opportunities relevant to the kinematics of the arm-camera system, without reliance on an intelligent teacher. In this section, we first define the standard single-layer supervised connectionist associative learning problem. Then we describe how a concern for biologically-plausible learning mechanisms has suggested a repackaging of a number of old ideas in the form of a new synaptic learning scheme, called *sigma-pi* learning.

Connectionist Supervised Associative Learning

Connectionism is the study of systems composed of large numbers of simple processors that communicate only through excitation and inhibition, and in which knowledge is stored in the strengths of the weighted interconnections between the simple processors [53]. We consider two populations X and Y of simple units (fig. 3.2). The instantaneous activity states of X and Y can be represented by the vectors $\mathbf{x} = \{x_1, \ldots, x_n\}$ and $\mathbf{y} = \{y_1, \ldots, y_m\}$, where the x_i and y_j represent the activity levels of the ith and jth units in populations X and Y, respectively. Population Y receives an "unconditioned" teacher input vector $\mathbf{t}$. In addition, X projects to Y via a single layer of "conditioned" or "associational" synapses that are modifiable. During learning, input population X moves through a sequence of states $\mathbf{x}^{(p)}$, $p = 1, \ldots, N$; for each, Y is "clamped" by its teacher input to an "associated" state $\mathbf{y}^{(p)} = \mathbf{t}^{(p)}$. The underlying relation

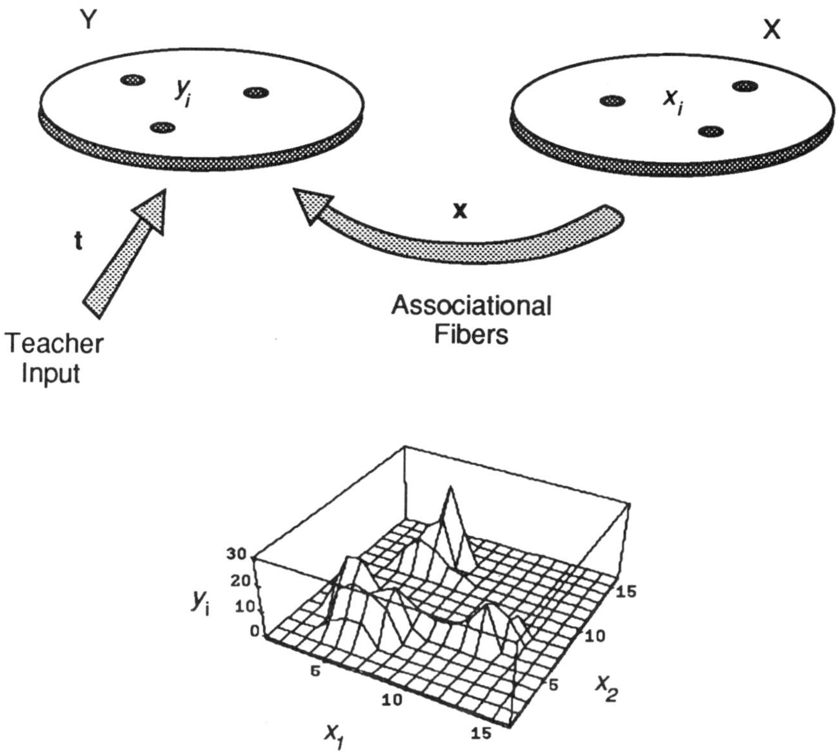

Figure 3.2: An instance of the standard connectionist associative learning problem. Input population X projects to output population Y via a modifiable associational pathway; during learning population Y is "clamped" by unconditioned teacher input vector $\mathbf{t} = f(\mathbf{x})$ as population X moves through its state space. The task of each individual output unit y_i is to learn its unconditioned teacher activation function $y_i = f_i(\mathbf{x})$ given the training data, which consists of a set of $(\mathbf{x}, y_i)$ pairs. In short, each of the y_i must learn to predict the intensity of its teacher signal solely on the basis of the conditioned input vector from X; the graph above illustrates an unconditioned teacher function plotted over two input variables. Function learning of this type can be viewed as the process of building up a complicated receptive field over the input space.

$\mathbf{y} = \mathbf{t} = f(\mathbf{x})$ is assumed to be a function, nonlinear in general, where the form of f is determined by the "semantics" of X and Y. The learning problem, then, is to build or modify synapses on the association pathway from X to Y such that each trained pattern $\mathbf{x}^{(p)}$ on X gives rise to a good approximation of $\mathbf{y}^{(p)} = f(\mathbf{x}^{(p)})$ on Y, typically one that minimizes mean-squared error over the training set. The learned map should also generalize such that novel $\mathbf{x}$ give rise to "reasonable" $\mathbf{y}$ (for discussion of this issue see [182, 77, 163]).

For example, we consider the case where the input population X is the joint-angle population and output population Y is the visual-field population. The teacher input $\mathbf{t}$ to population Y corresponds to the camera-derived input to the visual-field population. The learning problem from above may then be restated from the perspective of a single visual-field output neuron y. During training, y is occasionally fired by its camera input line t as the joint-angle population X moves through its 3-dimensional state space. If t is plotted over the joint space, the problem faced by visual-field unit y can be seen to be one of real-valued nonlinear function learning from examples in a 3-dimensional state space. Given the physics of the situation, these functions $\mathbf{y} = \mathbf{t} = f(\mathbf{x})$ over the joint space are generally sparse (i.e., mostly zero-valued), non-negative, and locally smooth. Since each visual-field unit has a different teacher input line, each must learn a different function.

The principal question remaining, then, concerns the way in which synapses are actually to be used and modified during learning. In the following, we sketch the historical progression in the study of synaptic learning rules, both from the biological and computational perspectives, and motivate a set of assumptions needed to strike a delicate balance between learning power and biological plausibility.

A Historical Quandary

In the past two decades, experimental neuroscientists have begun to work out the rules that govern synaptic plasticity during learning in the vertebrate central nervous system [27, 132, 122, 101, 7, 95]. In confirmation of Donald Hebb's 1949 conjecture [69], long-term strengthening of synapses that depends of simultaneous pre- and post-synaptic activation has been demonstrated in hippocampus and other cortical structures (for reviews see [32, 103]). Hebb hypothesized that such a rule would lead to the formation of mutually-excitatory "cell assemblies", which he argued could be a means for long-term storage of brain states and/or associations. The term "Hebb-type" will hereafter be used to describe any use-dependent rule for synapse modification that relies only upon locally-available pre- and post-synaptic activity levels [188]. While these recent experimental discoveries of Hebb-type learning rules are undisputed milestones in the study of neural learning mechanisms, it is sobering to note that synaptic learning rules of this simple type were all but abandoned by the theoretical learning community in the early 1970's, due to profound limitations in their ability to learn general-purpose associations among patterns [145].

This point is crucial and deserves elaboration. In contrast to the approach of experimental neurobiologists, the thrust of ongoing computational efforts in neural learning has been to explore a range of abstract "neural network" models that relate specific rules for synaptic weight modification to their consequences at the network level, such as their ability to encode patterns and associations of varying complexity and under varying conditions of supervision (for review of concepts see [145, 181, 77]). Hebb-type learning rules have been extensively studied in this way [110], most often for linear or pseudo-linear "neuron-like" units, and their scope and limitations are now well understood. They may generally be classified as "single-layer" learning rules, since they only specify changes in a single layer of synaptic weights between inputs and outputs. Single-layer learning mechanisms have proven useful for simple supervised pattern association problems [110], unsupervised clustering based on statistics of input patterns [67, 110, 184], and as the basis of certain developmental mechanisms, such as the establishment of receptive field properties in visual cortex [100, 123, 142, 150, 196, 213]. The notorious limitation of single-layer learning rules derives from the fact that they can capture only very simple functional relationships between a given pre-synaptic and post-synaptic unit, i.e., that which may be represented by a simple positive or negative "weight". The exclusive-or (XOR) function is often cited as an example of a very simple function that cannot be learned in this way, but is representative of *all* nonlinear functions in the sense that the influence of a given input on the output is not expressible as a fixed weight, but depends on the values of other inputs (e.g., consider $y = x_1 x_2$; the effect of positive x_1 on y can range from excitatory to none to inhibitory depending on the sign of x_2, and *vice versa.*) Thus, while single-layer learning rules are attractively simple from a biological perspective, and are now known to exist in the brain based on solid physiological evidence, they are, under usual assumptions, provably inadequate for general-purpose associative learning tasks of which the brain is presumed to be capable.

New Algorithms Are More Powerful, Less Biological

This limitation, first formalized by Minsky & Papert [145] for the case of Perceptron learning, has led to the development of more powerful abstract network learning schemes capable of modifying synaptic weights in *multi*-layered networks of pseudo-linear units (e.g., [2, 182]). The advantage of multiple layers of weights is that the effect of a given input unit on the output can in principle be made an arbitrary nonlinear function [120]. One example of such a learning scheme is *back-propagation*, a weight-space gradient-descent algorithm designed to minimize response errors over an input-output training set [182, 120, 157]. For each input-output pair presented to such a network (often called a multi-layered Perceptron or MLP), synaptic weights can in principle be modified throughout the entire multi-layer network, such that responses to all subsequent inputs are slightly altered. This property affords the MLP its power: what is learned from a single isolated input-output example contributes to the development of "hidden" global feature detectors that simplify the learning problem for all inputs [182]. The automatic construction of global features that simplify an input-output relation is an extremely exciting possibility, since it is a means for simultaneously reducing the dimensionality of the learning problem (making learning much faster), and providing for good generalizations to novel inputs.

Unfortunately, while the new generation of more powerful neural-network learning schemes have overcome certain limitations of their single-layer predecessors, they have introduced "anatomical" and "physiological" complexities that, coupled with poor scaling behavior and the fundamental problem of local minima, make them highly improbable as biological models [77]. The principal difficulty lies in the nature of the learning algorithm—a gradient-descent search through the parameter space of a high-dimensional nonlinear model[2]. Searches of this kind are well known to be expensive at best, and intractable at worst. Compounding the search problem is the fact that the space of synaptic weights is not fixed, since the number of internal units and layers needed to make learning work well is variable and strongly problem dependent. Moreover, when the relationship between inputs and outputs is complicated (i.e., nonlinear), the synaptic weight space to be searched becomes pocked with local minima that frustrate any "weak method", such as gradient descent. It has been observed often among neural network aficionados that one of the most potent engineering techniques in the case of a very difficult learning problem is to build in the best possible features from the start, in order that the gradient-descent learning process have the best chance of succeeding [200, 42]. Thus careful choice of input and output representations is, in many cases, a far more important ingredient in the success of neural learning than the sheer horsepower of the underlying synaptic modification scheme.

The second drawback of multi-layered synaptic learning schemes when considered as biological models concerns the significant complexities of implementation. In order to build internal feature detectors that help to fit input-output pairs seen during training, it is necessary that each synapse be modified based on its effect on all subsequent synapses and units on the path to the output layer. The rule for synaptic modification no longer depends on information available directly at the synapse, as it does for Hebb-type learning rules, but rather relies on information accumulated from units and synapses

[2]The MLP uses global geometric primitives to fit functions. The simplest hidden features, one layer removed from the inputs, are "squashed" hyperplanes situated over the input space. Since hyperplanes are defined globally (i.e., for all input vectors), perturbing any one of them can change system response globally. In this light, the learning problem becomes one of iteratively perturbing the set of initially random squashed hyperplanes until, as a set of global basis functions, they can be best used to fit the given set of input-output pairs.

across the network as a whole. In the case of back-propagation, this entails that derivatives of the response error with respect to each weight in the network be propagated backwards from outputs toward inputs. Extra circuitry must therefore exist for this purpose, or the existing circuitry must act in much more complex ways under global network control. In addition, each model synapse must be able to move freely through both excitatory and inhibitory values; since it is known with relative certainty that biological synapses are not capable of changing sign in this way, it becomes necessary to posit the existence of a network of "sign change" neurons. Unfortunately, this possibility too is unbiological [41]. Yet another possible stumbling block lies in the needed accuracy of the computations that underlie globally-generalizing multi-layer learning schemes, in relation to the small number of bits that may be available in single-neuron "arithmetic" hardware.

Changing Assumptions

Is it, then, a fundamental law that neural associative learning algorithms must be either representationally impoverished or mechanistically overcomplex? A good resolution to this quandary is achieved by upgrading several basic assumptions of neuronal processing that have run implicitly throughout the above discussion and much of the work discussed therein. First, we place strict representational conditions on the response properties of both input and output populations participating in the associative learning process, immediately simplifying the task of the synaptic modification rule; specifically, we assume coarse-coded (or "value-coded") populations as discussed in chapter 2. Second, we reject the standard summing-unit model neuron (dating to the McCulloch & Pitts neurons (1943), the Adaline [212], and the Perceptron [179]), in favor of the *sigma-pi* model neuron, which computes its output as a sum of independent multiplicative clusters of inputs. In so doing we provide a concrete scenario in which local interactions among inputs to a neuron, mostly ignored in current connectionist modeling efforts, are nontrivially utilized as an integral part of the learning computation. For reasons of computational efficacy, model neurons with multiplicative terms (often called "higher-order" units) have been studied by numerous workers (e.g., [18, 63, 135, 135, 46]); the potential biological utility of multiplicative, or AND-like, dendritic nonlinearities in real neurons has also been pointed out by numerous other authors [109, 193, 192, 108, 167]. Third, we propose a single-layer learning rule called *sigma-pi* learning, in which weights are modified for *clusters* of synapses, rather than for individual synapses, according to a standard Hebb-type learning rule. This general possibility of local interactions in synaptic learning, too, has been discussed elsewhere [58, 59, 33]. As will become apparent in the following, the foregoing assumptions lead directly to a learning scheme in which individual units act as locally-smoothing multi-dimensional lookup tables. We first discuss the concept of a lookup-table for function learning in the abstract, and then examine its neural relevance.

Learning Functions with Lookup-Tables

Perhaps the simplest way to learn a real-valued map $f(\mathbf{x}) : \mathbf{x} \mapsto y$ over a k-dimensional input space is to build a lookup table, i.e., to store function values y into a k-dimensional table of numbers indexed by the components of the input vector $\mathbf{x}$. Given an $(\mathbf{x}, y)$ pair to be learned, the $\mathbf{x}$ vector is used to index the appropriate table location (a k-dimensional box in the input space), and the associated y value is stored there either outright, or combined (e.g., averaged) with a previously stored value. As shown in fig. 3.3, this process of associating function values with localized regions of the input space is a means for approximating the graph of the function being learned. Learning with Radial Basis Functions (RBF's) is a closely related method [165], which has been recently reviewed and linked to regularization theory by Poggio & Girosi [163]. Rather than operate directly on the input space, input vectors are first "decoded" by a population of receptive field units with centers ξ_i, each of which represents a local, often radially-symmetric, region in the input space. Thus, an output unit computes its activation level as a weighted sum over a set of units with appropriately placed receptive fields, i.e., $y = \sum_i w_i g(x - \xi_i)$, where g defines a *radial basis function*, commonly a Gaussian, and w_i is its weight (fig. 3.4). The learning problem can then be charac-

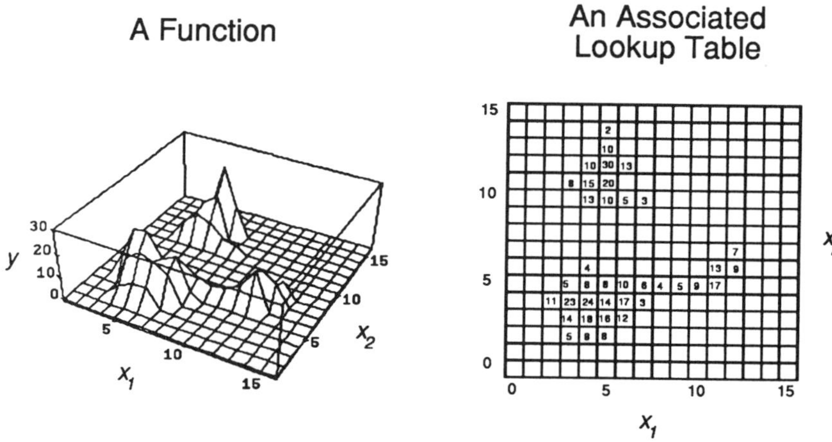

Figure 3.3: Graph of a function of two variables, and a corresponding lookup-table. Table entries are assumed to have value zero by default. From a neural perspective, table entries at right may be viewed as non-overlapping receptive fields in the input space.

terized as one of finding weights $\mathbf{w} = \{w_1, \ldots, w_n\}$ that minimize the squared difference between the actual function value and the one produced by the output unit in training, averaged over the set of training examples. Learning schemes of this type lend themselves directly to simple Hebb-type rules for synaptic modification since the initially nonlinear learning problem is transformed into a linear one in the unknown parameters $\mathbf{w}$ (see [31]). Atkeson [10] reviews a number of other table-based approaches to learning, and Omohundro [155] describes a range of efficient computer algorithms and data structures for their implementation, such as k-d trees (see pp. 48-50) and more recently "ball trees" [156]. A number of workers have explored connectionist versions of the locally-smoothing function lookup-table idea [4, 31, 53, 14, 117, 135, 136, 143, 149, 209].

The classical advantages of tabular learning schemes, neural or otherwise, are learning speed, algorithmic simplicity, and lack of strong representational bias (i.e., *a priori* preference for a specific class of input-output functions). Basically, the two requirements for a tabular approach to learning are:

i) a means for detecting specific regions of the input space, and

ii) a means for associating and/or retrieving a value for each such region.

In a neural context, these operations may be interpreted as:

i) establishing a set of appropriately placed receptive fields in the input space, and

ii) attaching a weight to each receptive field given by the desired function value to be associated with that region.

Figure 3.4 depicts a neural "lookup-table", encoding a nonlinear function over a two-dimensional input space. Receptive fields are distributed over the input space only where the function has nonzero value, and are weighted according to the average height of the function within their respective receptive fields.

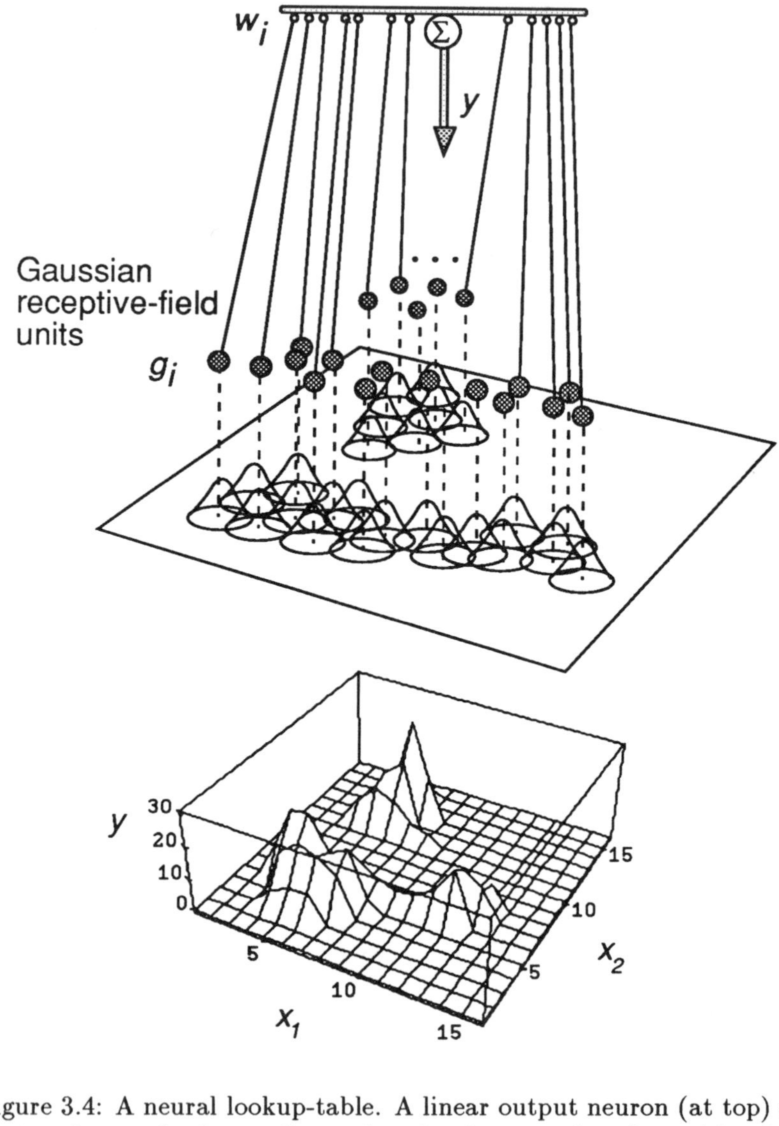

Figure 3.4: A neural lookup-table. A linear output neuron (at top) approximates the graph of a nonlinear function by summing the weighted outputs of a set of localized receptive field units distributed across the 2-dimensional input space. Where the function is zero-valued, no receptive field units are needed. Also, fewer (broader) receptive field units are needed, in general, when the function to be learned is relatively smooth.

The classical disadvantages of tabular learning derive from the exponential time/space requirements associated with the construction of a multi-dimensional lookup table[3]. However, several effects can mitigate the problem of exponential blowup. First, trivially, the functions of interest may be of relatively low input dimension such that the lookup-table can be of tractable size. Second, the functions to be learned may be sparsely active, i.e., mostly zero-valued, such that table space may be substantially conserved using standard representational techniques; in the neural analogy of fig. 3.4, we simply omit receptive fields everywhere the function is zero-valued. Third, the functions to be learned may be smooth (even though nonlinear), such that fewer table entries (or receptive fields) are needed to achieve a given accuracy of local approximation. Fourth, large portions of the multi-dimensional input space may be of little or no use in practice. These regions may therefore be considered as "don't cares", further reducing the space/time requirements of learning. All four of these mitigating factors hold for MURPHY's visual-kinematic maps, and are likely to hold in a wide variety of other sensory-motor learning problems of biological relevance.

Building Receptive Fields With Multiplication

We have seen that a lookup-table has a natural biological implementation given the ability to construct, place, and weight multi-dimensional receptive fields in the input space. A simple fact that has often been noted in the literature (e.g., [14, 135, 163]) is that higher-dimensional receptive fields can be constructed by multiplying together the outputs of units with lower-dimensional receptive fields. For example, fig. 3.5 shows that a 2-dimensional Gaussian results from multiplying two one-dimensional Gaussians. In a sensory context, this implies that a neuron selective simultaneously for stimulus parameters along two dimensions, such as the angles of two joints conjunctively, could be derived by multiplying together two

[3]The worst-case storage requirements occur for functions that vary erratically across the entire input space, since this case requires that an exponentially large lookup-table be completely filled (storage requirements are $O(g^k)$, where g is the number of table gradations per input dimension, defining the "grain" of the approximation, and k is the number of input dimensions.)

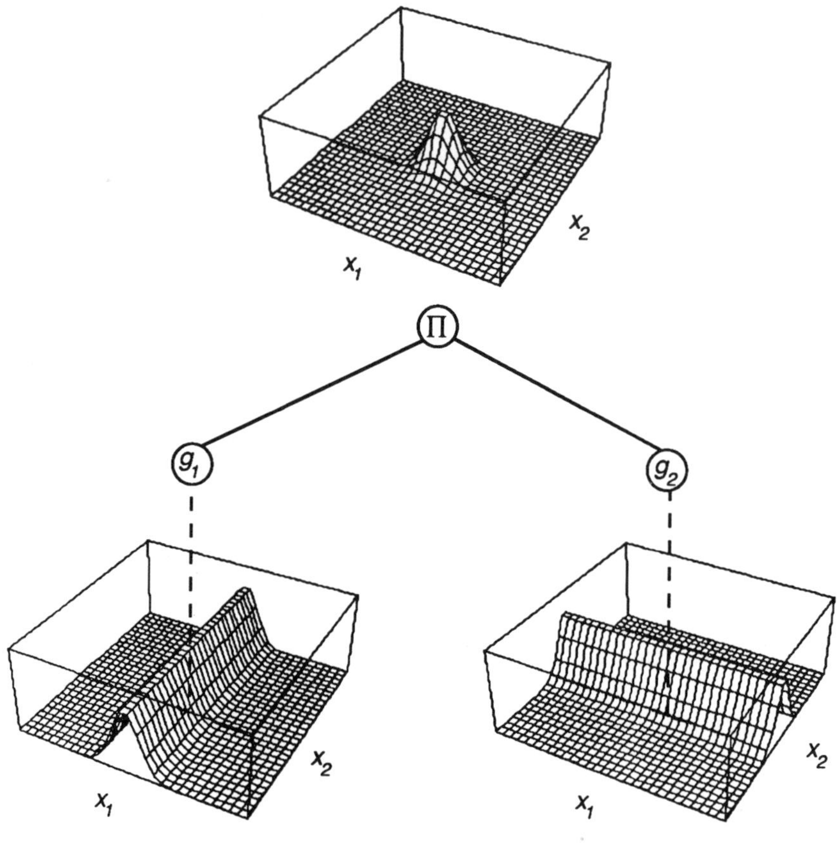

Figure 3.5: A 2-dimensional Gaussian receptive field may be constructed by multiplying together the outputs of two units with one-dimensional Gaussian receptive fields. Multi-dimensional neural response selectivities can be synthesized in this way. For example, a unit simultaneously selective for the positions of two joints can be constructed by multiplying together the outputs of two units, each of which is selective for the position of only one of the joints.

inputs, each selective for the angle of only one of the joints. For a detailed review of these issues from a neurobiological perspective, see [138].

Having motivated multiplication as an elemental input operation for neural learning, we now introduce the *sigma-pi* unit, and relate it to the lookup-table abstraction discussed above.

Sigma-Pi Learning

The Sigma-Pi Unit

Unlike the classic thresholded linear unit that is the mainstay of many current connectionist models, the output y of a *sigma-pi* unit is computed as a sum of contributions from a set of independent multiplicative clusters of input weights:

$$y = \sum_j w_j c_j, \tag{3.1}$$

where

$$c_j = \prod_{i=1}^{k} v_i x_i \tag{3.2}$$

is the product of k inputs x_i with their weights v_i within cluster j, and w_j is the weight on cluster j as a whole. During learning, the output may be clamped by a teacher input, $y = t$ (fig. 3.6).

Units of this general type were first proposed by Feldman & Ballard [53], and have been used occasionally by other connectionist modelers, most commonly to allow certain inputs to gate others or to allow the activation of one unit to control the strength of interconnection between two other units [181]. The use of *sigma-pi* units as function lookup-tables was suggested by Feldman & Ballard [53], who suggested a possible relevance to local dendritic interactions among synaptic inputs. Volper & Hampson [207] have also discussed the use of related OR-of-AND neurons as well as neurons with polynomial inputs [208] for the learning of Boolean functions from examples, and have made certain ties to properties of biological learning. *Sigma-pi* units are also often described as "higher order" units, since the multiplicative operation can be used to implement

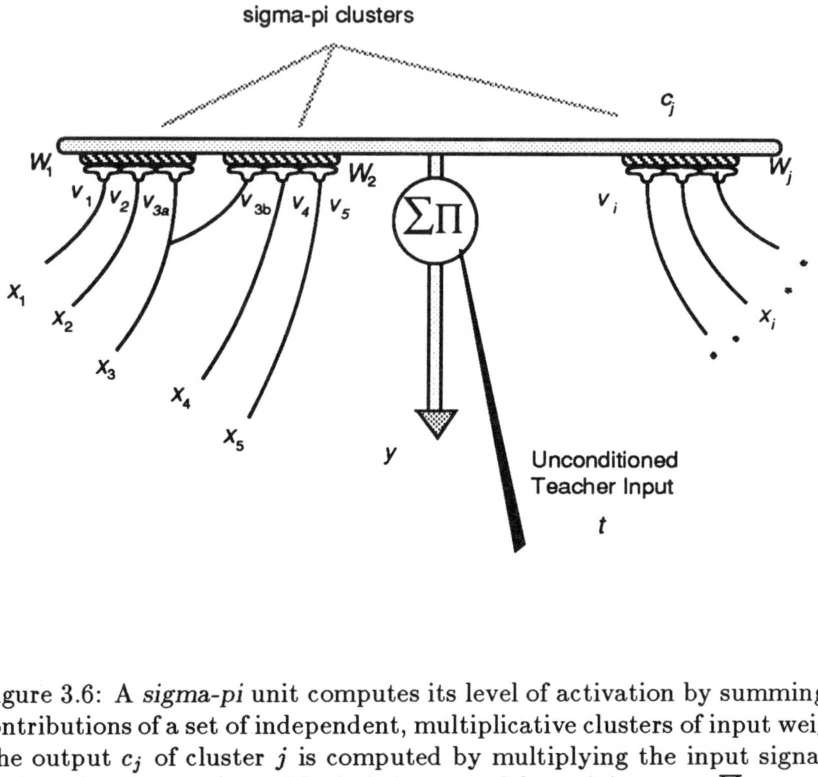

Figure 3.6: A *sigma-pi* unit computes its level of activation by summing the contributions of a set of independent, multiplicative clusters of input weights. The output c_j of cluster j is computed by multiplying the input signals x_i to that cluster together with their input weights, giving $c_j = \prod_i v_i x_i$. The output y of the entire unit is computed as a weighted sum of the outputs of each cluster, giving $y = \sum_j w_j c_j$. In the current context, a *sigma-pi* cluster may be viewed as a multi-dimensional receptive field centered on a specific input pattern.

global second- and higher-order polynomial relations among a set of inputs [63]. Durbin & Rumelhart [46] have recently used related "product" units in an extension of the back-propagation algorithm to learn arbitrary polynomial functions over the input space; they too suggest possible relevance of their multiplicative nonlinearity to local processing operations in dendritic trees.

In the present work, the specific nonlinear interaction among inputs to a *sigma-pi* cluster is not of primary theoretical importance. The crucial property of a cluster is that its output should be AND-like, i.e., selective for the simultaneous activity of *all* of its k input lines.

The Learning Rule

We begin by assuming all possible clusters of size $1 < k < d_{\mathcal{X}}$ pre-exist with zero weight, where $d_{\mathcal{X}}$ is the dimension of the underlying input space. In the case of MURPHY's forward kinematic map, $d_{\mathcal{X}} = 3$, i.e., the dimension of the space of 3 joints. The cluster weights w_j are modified during training according to the following self-normalizing Hebb rule:

$$\dot{w}_j = \alpha c_j t_j - \beta c_j w_j, \tag{3.3}$$

where α and β are small positive constants, c_j is the cluster response, and t_j is the teacher input to cluster j. Input weights v_i are of unit size and are unmodified during learning. The steady state of this learning rule occurs when

$$w_j = \frac{\alpha}{\beta} \frac{<c_j t>}{<c_j>}, \tag{3.4}$$

i.e., when w_j converges, very roughly speaking, to the expected value of the teacher signal within the cluster's receptive field, normalized by the size of the receptive field. Equation 3.3 includes both a reward term as well as a decay term, and tries to maximize the correlation (strictly speaking, the average product) of cluster activation and the teacher signal over the training set, while minimizing the total weight for all clusters (weighted in turn by their average levels of activation)[4]. Motivation for the second, decay term in equation 3.3 is most easily seen in the lookup-table analogy for function approximation. The weight given to each cluster receptive field should be proportional to the expected intensity of the teacher signal *when in that cluster's receptive field.* To this end, the decay term in equation 3.3 is proportional to the cluster's activity c_j, insuring that those clusters are relatively "punished" that fire too often without reinforcement from the post-synaptic teacher signal (i.e., for whom the expected intensity of teacher reinforcement is low). This weakening condition, for sparsely-active functions, can be shown both analytically and in simulations to punish clusters of synapses that are either i) non-selective, such as a group of synapses that *always* fire, independently of the teacher signal, ii) useless, such as a group of synapses that never fire, or iii) poor predictors, such as a group of synapses that rarely fire together in conjunction with the post-synaptic teacher signal.

An Example

We now complete our example of the learning of MURPHY's forward kinematic model. (Note that the learning of the inverse differential map proceeds in a completely analogous fashion). The camera-derived teacher function t for a given visual-field unit y during training is a nonlinear function over the 3-dimensional joint space, whose graph is determined by the kinematics of the arm, the location of visual features (i.e., white spots) on the arm, the global properties of the camera/imaging system, and the location of the visual unit's receptive field (fig. 3.7).

We recall that MURPHY's joint-angle population is coarse-coded by dimension, such that each unit's response field is centered on a

[4]More formally, the learning rule can be seen to minimize the cost functional,

$$E = -\sum_{p}(\alpha t^{(p)} y^{(p)} - \frac{\beta}{2}\sum_{j} c_j^{(p)} w_j^{(p)2}),$$

where the first term grows as the product of the desired and actual output of the unit, and the second, punitive, term tends to limit the total weight allocated to all clusters, where a larger cost is accrued for clusters that are active more frequently.

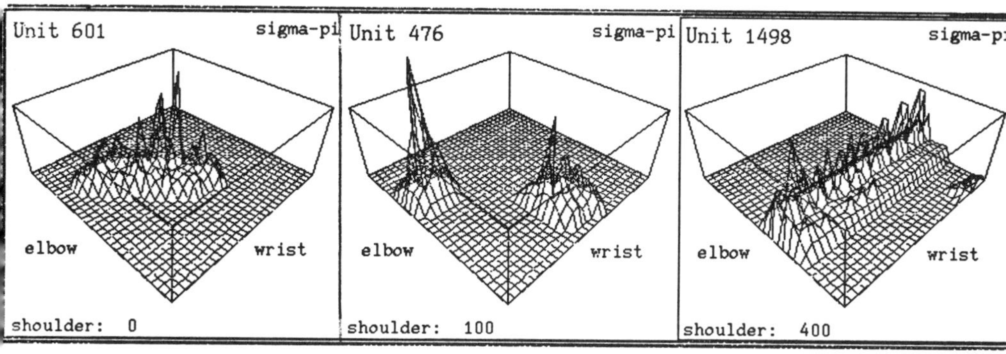

Figure 3.7: Plots of the *sigma-pi* approximations to the camera-driven "teacher" functions for three visual-field units. Plots are over two of the three joint variables.

specific angular value for one of the three joints. It was pointed out (pp. 15-16) that in this representation, a global arm configuration can be approximately encoded as the *conjunctive* activity of the three most active joint units, i.e., the "loudest" unit in each of the three subpopulations. (Note that this code is only approximate since the input state will be quantized to the nearest represented joint angle peak for each of the joints. This fact results in pressure for small steps in the joint-angle peaks from unit to unit.) The process of *sigma-pi* cluster formation for a single visual-field unit is illustrated schematically in fig. 3.8 as a one-shot process. The arm is assumed to be in configuration C, with wrist = 60°, elbow = 30°, and shoulder = 100°. As shown in the graph of fig. 3.8, visual unit V_{476} is fired in this configuration by its camera input with intensity $t = 80$. The three maximally-activated joint units J^w_{60}, J^e_{30}, J^s_{100} (i.e., with peaks at these joint angles) form a new *sigma-pi* cluster at the input to V_{476}, with input weights v_i of unit size and cluster weight $w \propto t = 80$. (In practice, this weight can be slowly adapted to its final value according to equation 3.3.) After learning, V_{476} will be activated by this cluster *predictively* whenever the arm is "felt" to

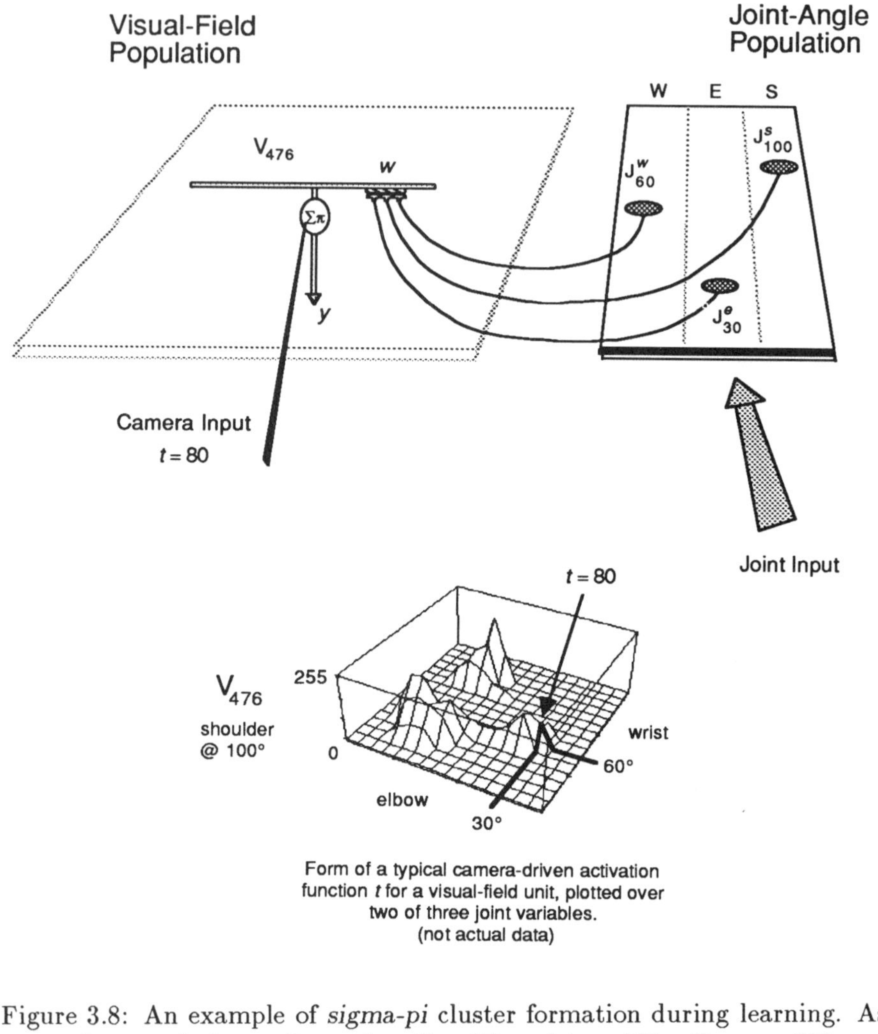

Figure 3.8: An example of *sigma-pi* cluster formation during learning. Assume the arm is in configuration C, with wrist = 60°, elbow = 30°, and shoulder = 100°. According to the graph above (not actual data) visual unit V_{476} is fired in this configuration by the camera input line with intensity $t \simeq 80$. The three maximally activated joint units J^w_{60}, J^e_{30}, J^s_{100} are recruited, in a Hebbian sense, to form a new *sigma-pi* cluster (i.e., table entry) at the input to V_{476}, with input weights of unit size and cluster weight w proportional to $t = 80$. During subsequent experience, V_{476} will be activated by this cluster *predictively* whenever the arm is "felt" to be in configuration C, signaling the imminent arrival of a white spot in its visual receptive field.

be in configuration C, which signals the imminent arrival of a white spot in its visual receptive field. This process of cluster formation is repeated for all other active visual-field units in each training state, with cluster weights w correspondingly scaled to each unit's level of camera-driven activation. Thus, each visual-field unit activated in configuration C simply records a "code" for that arm configuration as a three-element *sigma-pi* cluster. In other words, each multiplicative cluster acts essentially as a receptive field selective for a particular joint configuration. When again in configuration C, all of these visual-field units will be fired by their respective *sigma-pi* clusters, collectively producing a visual "mental image" of the arm in configuration C.

Generalization to Novel Inputs

Were it not possible to generalize to novel inputs, the practicality of this type of learning scheme would be greatly reduced. However, the neural lookup-table analogy discussed above suggests that good generalization is indeed possible. When, for example, a visual unit V is activated in a joint configuration C, and acquires an appropriate cluster of weights from the joint-angle population, by continuity V will generally be *partially* activated in *similar* joint configurations as well. Since the joint-angle population is coarsely tuned, generalization to similar joint configurations occurs automatically: as the arm is moved a small distance from a specifically-trained joint configuration, the output of the synaptic cluster encoding that training configuration will fall smoothly from its maximum, i.e., will contribute partial activation to its parent unit that decays gradually as the arm moves further and further from the trained configuration. In this way, a visual-field unit can fire predictively in joint states that it has never seen during training, by summing across the set of partially-activated clusters that encode *nearby* input states. More generally, each *sigma-pi* cluster can be described as representing a hyper-rectangular volume (or receptive field) in the 3-dimensional joint space over which it contributes activation to its parent unit; when training is sufficiently dense in the input space that the neighboring cluster volumes overlap, a visual-field unit can closely approximate its teacher function over the entire joint space. The exact functional form of the approximation depends on several system parameters, primarily the density and receptive field characteristics of the input units (i.e., the shape, radii, and degree of receptive-field overlap in the input population). A more precise analysis of this type of function approximation is underway [137], but for present pragmatic purposes, it suffices that the approximation be reasonably smooth.

In summary, *sigma-pi* learning is a scheme in which each neuron acts as a locally-smoothing multi-dimensional lookup-table whose entries are stored as distinct clusters of synapses. A lookup-table approach to associative learning is fast, simple to implement, very general, and of arguable biological relevance [138].

K-d Tree Reimplementation

Under the constraint of forced sequentiality (i.e., no parallelism available), the *sigma-pi* learning scheme is a poor choice for implementation. MURPHY develops 2.5 million weights during his 5-hour training regime; during a single iteration through the forward kinematic model, each of these weights must be accessed, multiplied, and accumulated in sequence. Whereas in its abstract specification a single run through the kinematic map may be defined as a single parallel machine operation, it is transformed into millions of machine operations on a serial computer. When each *sigma-pi* unit is viewed simply as a nonlinear function approximator, it may be observed that much more efficient algorithms are available for their implementation on serial computers.

Omohundro [155] reviews a wide range of data structures useful for learning functions from examples, with particular emphasis on multi-dimensional binary trees called "k-d trees", first described by Bentley [24]. K-d trees are a type of adaptive, multi-dimensional lookup-table that builds up the graph of a function by associating function values (or linear or higher-order approximating surfaces) with arbitrary hyper-rectangular regions of the input space. Each node in the binary decision tree represents a cut at a specified value along one of the input dimensions. Given an input vector for which

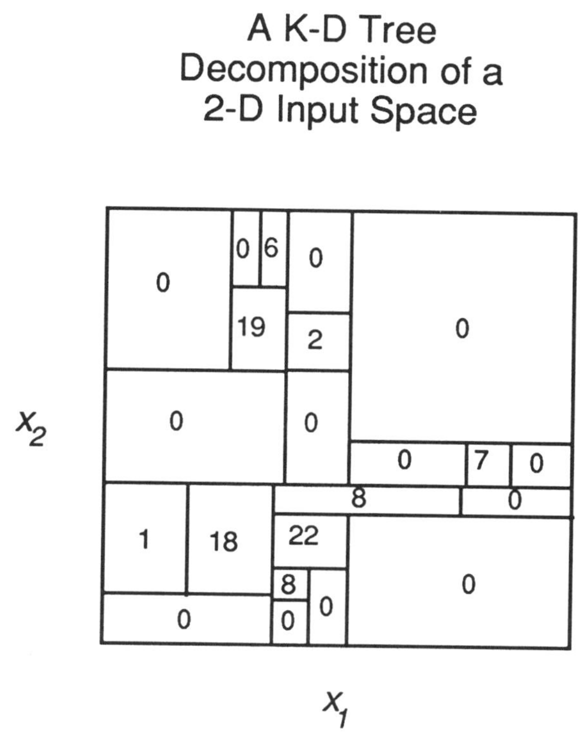

Figure 3.9: A k-d tree is a binary tree that can be used to partition a multi-dimensional input space for function approximation in a way that allows highly efficient implementation on a serial computer. For example, the tree of axis-parallel cuts above decomposes the input space from fig. 3.3 recursively into a set of non-overlapping hyper-rectangular regions (leaf nodes). Within each region, a locally-valid approximation to the function is stored, such as an average value (as pictured), or a linear or higher-order approximating surface. The k-d trees used to approximate MURPHY's forward kinematic map used constant approximators within the leaf nodes (see fig. 3.1), giving rise to a piecewise-constant approximation to the camera-driven teacher functions.

a function value is desired, the tree is descended as follows: at each node, the stored decision value v for dimension d is compared to the dth component x_d of the input vector $\mathbf{x}$. If $x_d v$, the tree is descended to the right, otherwise to the left. Leaf nodes represent non-overlapping hyper-rectangular regions in the input space, and contain the function values or higher-order approximating surfaces appropriate to their respective regions. The tree is constructed by recursively chopping the current region into two subregions with the best possible axis-parallel cut, i.e., such that the best-fitting approximating surfaces on both sides of the cut yield the minimum sum of squared errors across both subregions. The process is continued until the total error within each region falls below some threshold. The advantage of this type of representation is that lookup of a function value requires on average $O(\log n)$ steps, where n is the number of stored exemplars, i.e., the time needed to descend a binary tree with n leaf nodes. Post training therefore, MURPHY's 4,096 visual-field neurons and their weights are each replaced by a k-d tree, affording a speedup of $O(n/\log n)$ over the neural implementation, which in MURPHY's case amounts to a factor of between 50 and 100. Building the k-d trees takes 16 hours on a SUN 3-160. As can be seen in fig. 3.1, the k-d tree mode of operation produces mental images of the arm (lower right) that are essentially indistinguishable from those driven by the camera (upper right).

Though the k-d tree approach outlined here was sufficiently fast as to allow for convenient experimentation with MURPHY, it is very likely that other techniques for function approximation could result in still higher runtime performance (e.g., see [156, 124]).

Chapter 4

MURPHY in Action

In preceding sections we have seen the basic architecture and learning rule that allows MURPHY to build a "mental model" of his arm. To this point, we understand how MURPHY can generate a mental image of his arm in an arbitrary joint configuration that is sufficiently close to the real thing as to be useful for reasoning about visual goals. We now demonstrate that with only a handful of simple visual heuristics, MURPHY is able to solve difficult visually-guided reaching problems in the presence of obstacles.

Planning with Gradient Descent

Once MURPHY has learned to image his arm in an arbitrary joint configuration, he can use heuristic search to guide his arm "mentally" to a visually-specified target. In his simplest programmed reaching mode illustrated in fig. 4.1, MURPHY performs simple gradient descent in distance to the target. Each move is generated randomly, but only those that bring the hand closer to the target are saved. When the mental image of the hand is seen to fall in superposition with the visual target (within tolerances), MURPHY sends the arm physically to the final joint configuration in a single step, without concern for the path of movement. This reaching strategy has the advantage of being extremely simple to implement, but fails in general for any reaching problem that has local minima Local minima (i.e., dead-ends) arise when the arm either becomes tangled up in

Figure 4.1: Flailing toward a visual target. Left frame shows MURPHY in his initial configuration relative to a target (white cross). The center frame shows the trajectory of the hand during the sequence of "mental" steps taken as MURPHY searches for a path to the target. Each step is generated randomly; only those that reduce the distance to the target are saved. Once a goal configuration is found, the arm is moved physically in a single step (right frame).

its own joints, or with obstacles, such that no small movement can reduce the distance to the goal. In this case it becomes necessary to back up the arm, or the internal image of the arm, and try more promising avenues.

It is interesting to note that it may be difficult or impossible to differentiate, based on outward behavior, a system such as MURPHY that uses simple heuristic search on a forward-kinematic model of its arm from one relying on an explicit inverse kinematic model. Yet the difference is fundamental: in the traditional task decomposition in robot control, the problem of sending the tip of a robot arm to a desired location in workspace has been viewed as fundamentally a problem of inverse kinematics, an approach that intrinsically ignores the path of the arm. Obviously, this view is only valid in the case where no obstacles are present in the workspace; in the more general case, explicit path-planning algorithms are needed (e.g., [126]). In the current formulation, moving the arm to a desired location is always considered to be a problem of path planning, *regardless of the presence or absence of obstacles.* A very simple motion planner can therefore subsume the traditional function of an inverse kinematic model. In section 7 we argue this tradeoff is highly desirable from the perspective of biological implementation.

Motion Planning with Obstacles

Four Visual Routines

Simple heuristic search can also be used to plan movements for a multi-link arm in the presence of obstacles, even when the needed paths are complex and traverse much of the configuration space. For the simple case of reaching for an unobstructed target, MURPHY needs only to locate his hand in the visual field and to compute the straight-line distance from his hand to the target. For the more complex case with obstacles, MURPHY is endowed with the following additional "visual routines" (following Ullman [205]). First, on the basis of their size, MURPHY can label the blobs in the visual field as either targets or obstacles. Any convenient visual cues can be used for this discrimination, such as size, shape, or color. Second, MURPHY can identify his hand, wrist, elbow, and shoulder, also by their visual-field characteristics. The hand is distinguished trivially as a dual-blob (since there are two white spots on MURPHY's hand and only one on the wrist and elbow). The wrist and elbow spots are distinguished by their locations relative to the hand. The shoulder is always in a known visual-field location since it is fixed relative to the camera. This joint identification step is needed in order to allow explicit reconstruction of the "invisible" link shapes lying between the each of the identified joints. In the current implementation, straight lines were used as crude but easy-to-compute approximations of the link shapes (fig. 2.3). This geometric simplification is the explanation for the apparent collisions seen in figs. 4.2 to 4.9. Were the arm made completely visible to MURPHY in the video image during learning, this identification step would be unnecessary since the full image of the arm would be available at the output of the forward kinematic map, and collisions would thus be automatically avoided for the arm's entire volume. As discussed in chapter 2, however, serial emulation times for MURPHY's parallel algorithms grow in proportion to the amount of visual-field activity, resulting in pressure to keep visual activity to a minimum.

MURPHY's third visual routine involves the measurement of the distance from his hand to the target in a slightly more sophisticated manner than before. When there is no obstacle in the line of sight from hand to target, he computes the straight-line distance. If an obstacle lies in the line of sight to the target, then he measures distance to the target *circumventing* the first intersected obstacle to one side or the other, whichever results in a shorter total path. The resulting hinged path is only a very crude measure of remaining distance to the target, since it may cut straight through one or more subsequent obstacles, and may in fact represent entirely the wrong path. It is therefore somewhat surprising how well this distance heuristic functioned during planning, as is illustrated below. MURPHY's fourth and final visual routine entails two facts about collisions. If any of the joints or the explicitly computed link volumes visually overlaps with one of the obstacles, or if any part of the arm passes through any one of the obstacles during a mental step of the arm, then a collision has occurred.

Armed with this simple battery of visual routines, MURPHY follows these steps in reaching for a visual target:

The Search Procedure

1. Label the target and any obstacles present in the visual field.

2. Generate a mental image of the arm in the current joint configuration. Identify and label any parts of the arm that are visible in the mental image. (Some may be out of view.)

3. Measure the distance from hand to target, taking into account any intervening obstacles.

4. Mentally move the arm in a random direction (but remaining on a quantized grid in joint configuration space). This involves

randomly perturbing the internal joint-angle representation, and regenerating the mental image of the arm—all without physically moving the arm. If this arm configuration either i) has been tried before, ii) pushes the hand out of the field of view, or iii) causes a collision with an obstacle, then reject it and try another move. If there are no more moves to try from this position, go back a step and try the next-most-promising move. If all paths are exhausted, give up.

5. Measure the new distance to the goal. If it has decreased, loop to 4 and keep moving closer. If it has increased, temporarily reject it, i.e., push it on the stack of things to try later, and go back to 4 for another try.

6. If and when the hand reaches the target, move the arm physically to the target following the stored path, dispensing with all backtracking dead-ends.

Figures 4.2 through 4.9 demonstrate MURPHY's solutions to several reaching problems. In figs. 4.2 through 4.4, the first frame in the top row shows the arm in its initial configuration relative to a visual target (white cross) and obstacles. The second frame shows the sequence of intermediate hand positions generated during the mental search process. Once a complete mental trajectory has been found, MURPHY moves his hand physically to the target along a path that has been pruned of backtracking dead-ends encountered during the search process (third frame). The lower panels show a stop-action "movie" of MURPHY's solution path in the physical execution of the plan, read from left to right by row. Occasional collisions of "invisible" portions of the arm are exemplified in frames 14 through 17 in fig. 4.4; as discussed in chapter 2, these collisions are explained by the reduced internal image of the arm adopted as an efficiency compromise appropriate for the current serial implementation. The problem of fig. 4.3 is particularly difficult since the arm must be unwound in a cramped space before the target may be acquired, involving a complete traversal of the arm's configuration space and several deep local minima. In the process of unwinding, MURPHY is temporarily lured

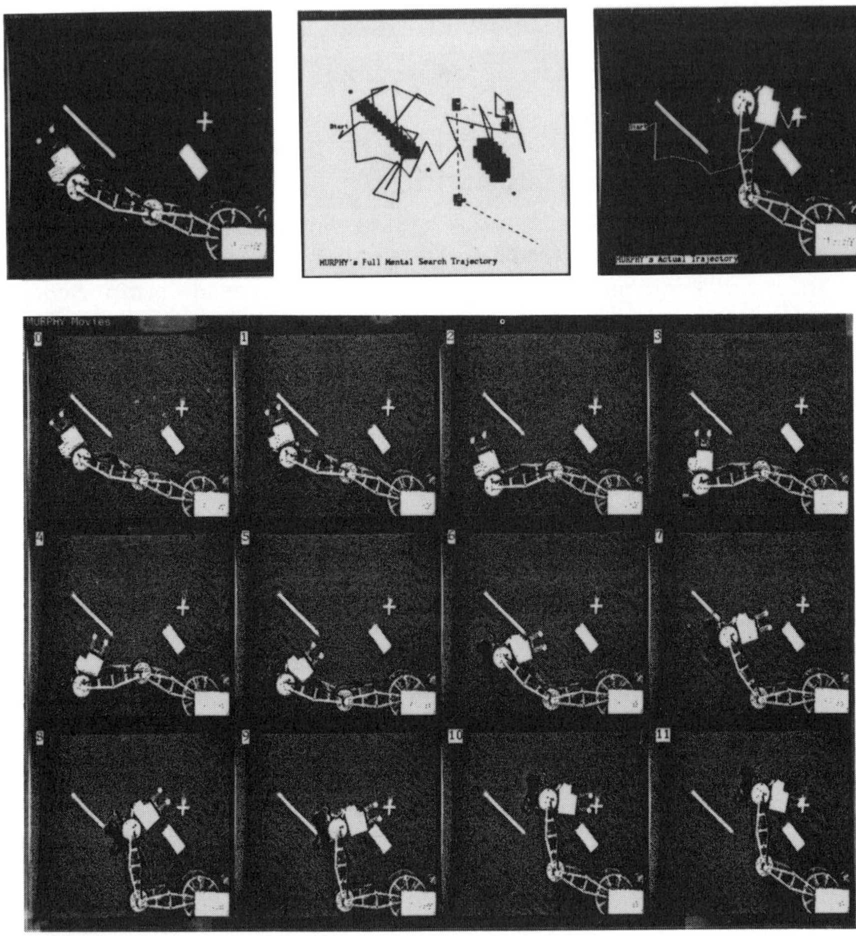

Figure 4.2: Reaching for a target in the presence of obstacles. In top row, left frame shows initial setup including target (white cross) and obstacles (other white shapes). Center frame shows trajectory of hand during mental search, and right frame shows hand trajectory on path to target. Lower panel shows the physical solution path once planning is complete.

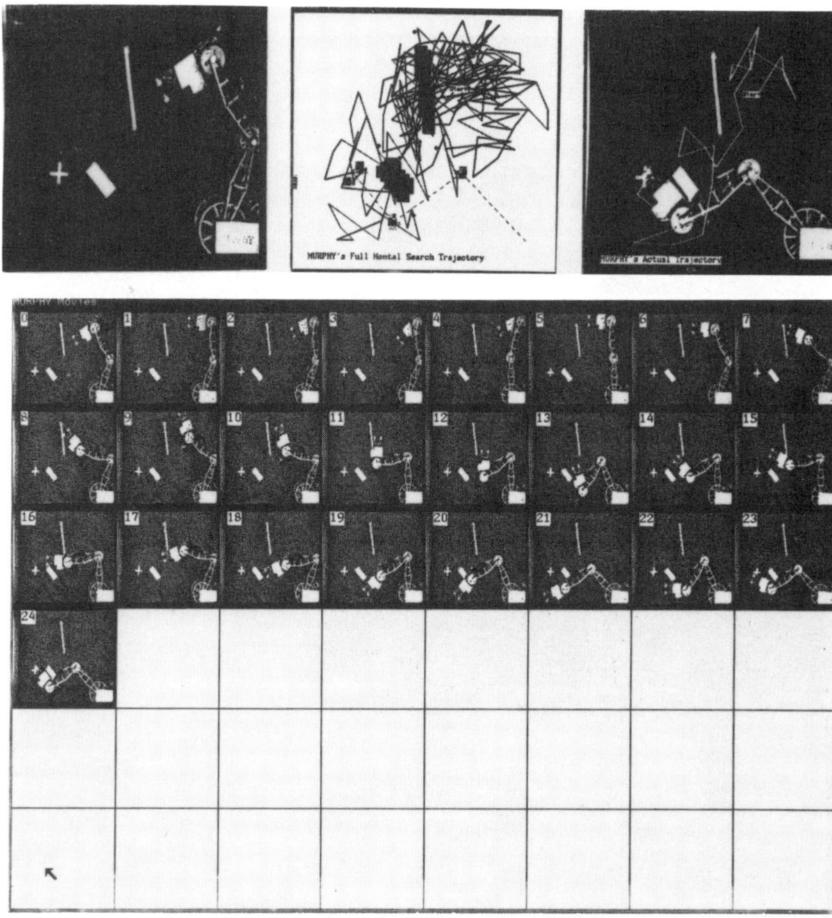

Figure 4.3: See text on p. 55 for general explanation. Frames 1 through 6 show how MURPHY's plans can be unnecessarily circuitous. Also note the unsuccessful attempt to reach over the lower obstacle in frames 14 through 16. Such false starts can remain in the final solution trajectory due to the nature of the best-first-at-every-depth search procedure.

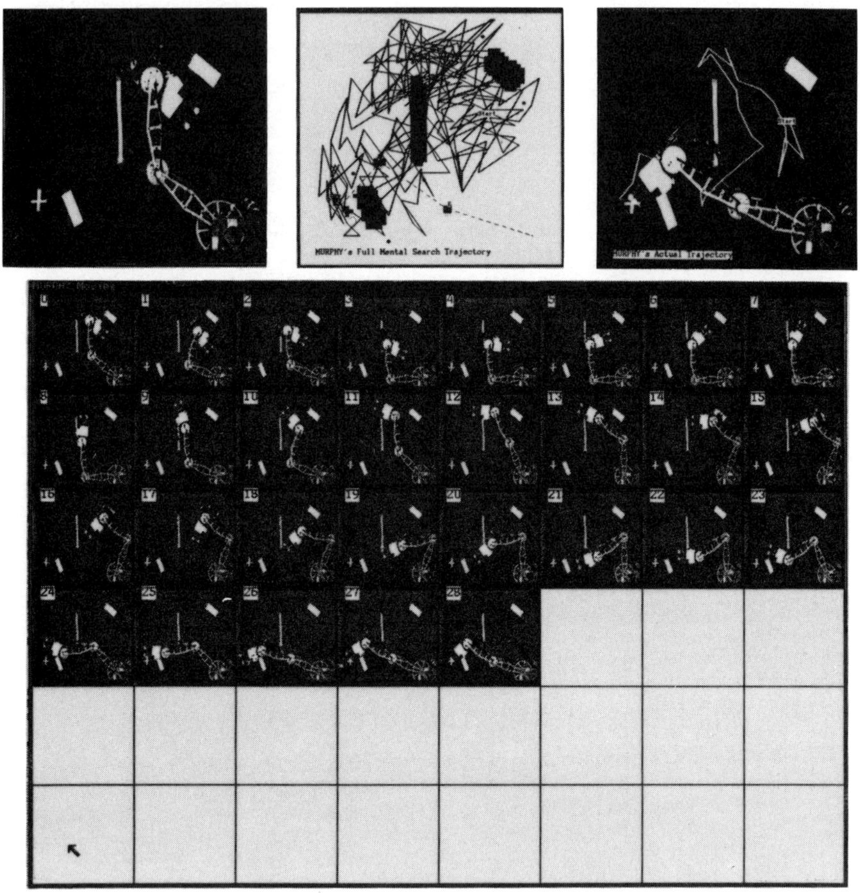

Figure 4.4: See text on p. 55 for general explanation. This reaching problem is very difficult, requiring that MURPHY completely unwind all three joints in a cramped space. However, a solution was found after only slightly more than 100 internal search steps.

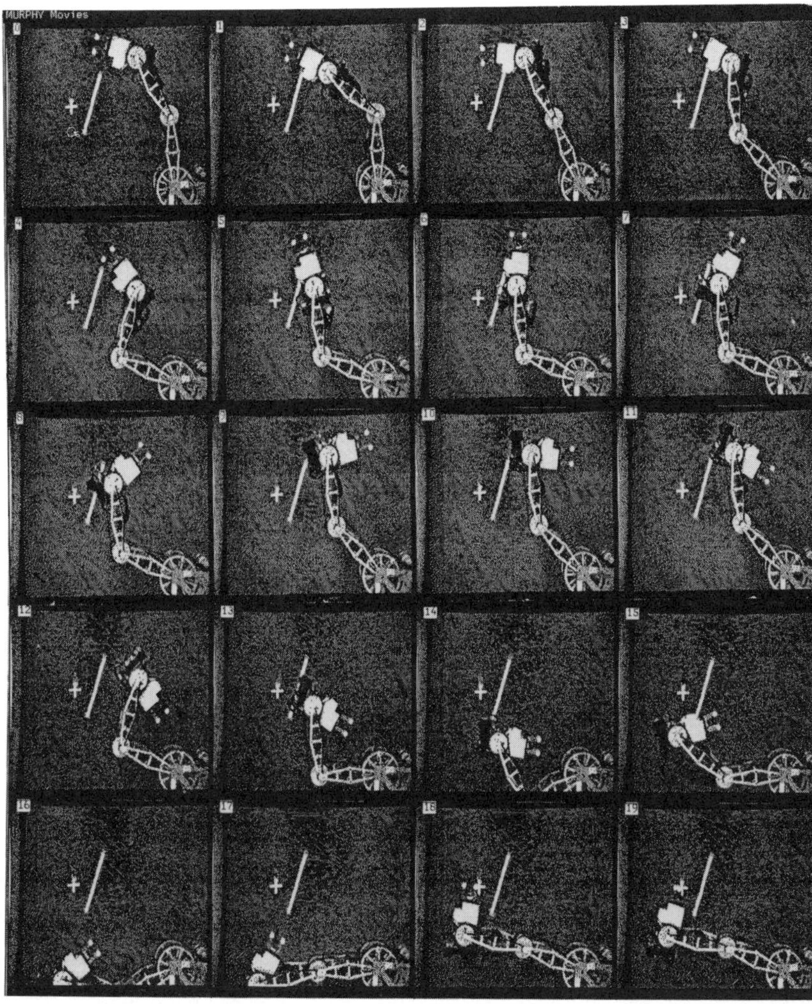

Figure 4.5: MURPHY's reach for an obstructed visual target. Note the apparent collision in frame 13 due to the imperfect geometric model of the arm that is used to explicitly fill in the invisible arm parts in the mental image. This problem can be easily corrected either with a more realistic explicit model of the arm's invisible regions, or by using a more complete visual representation of the arm during learning.

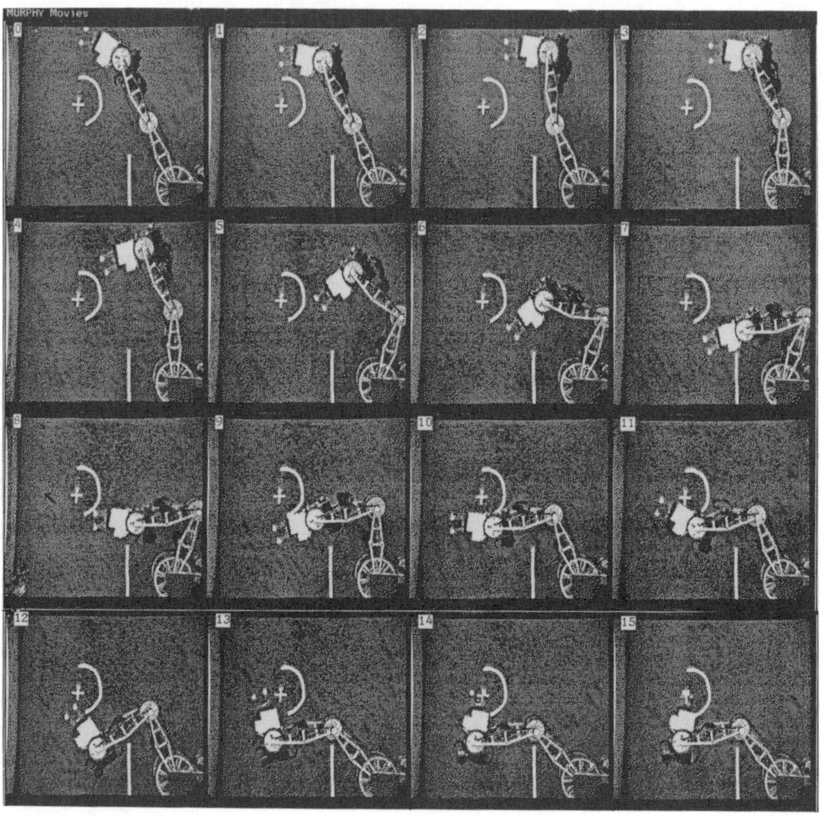

Figure 4.6: MURPHY's reach for an obstructed visual target.

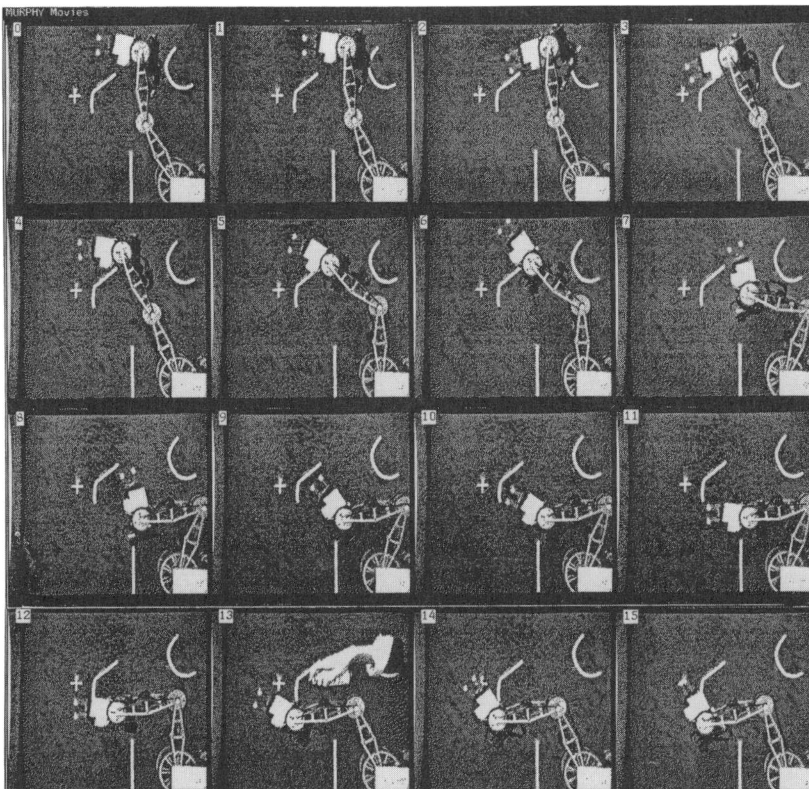

Figure 4.7: MURPHY's reach for an obstructed visual target. Note how the addition of a single obstacle to the workspace from fig. 4.6 significantly alters MURPHY's reaching solution.

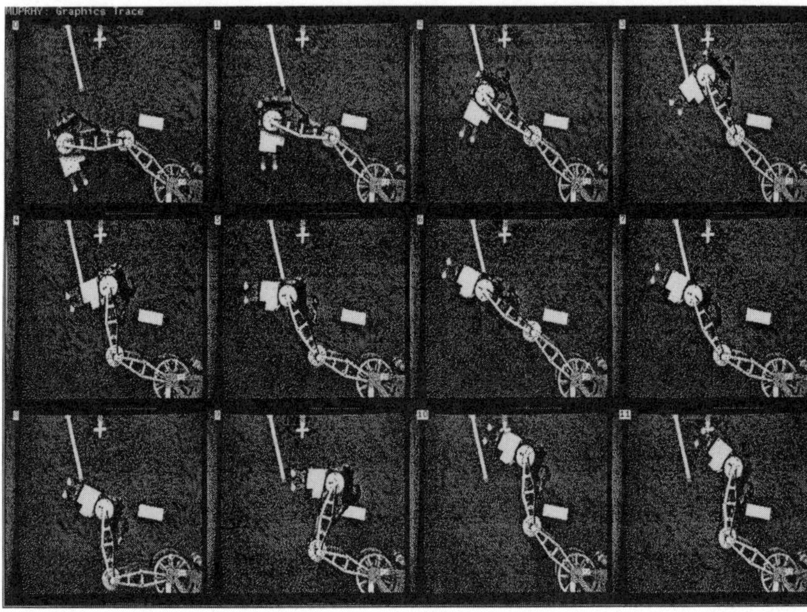

Figure 4.8: MURPHY's reach for an obstructed visual target.

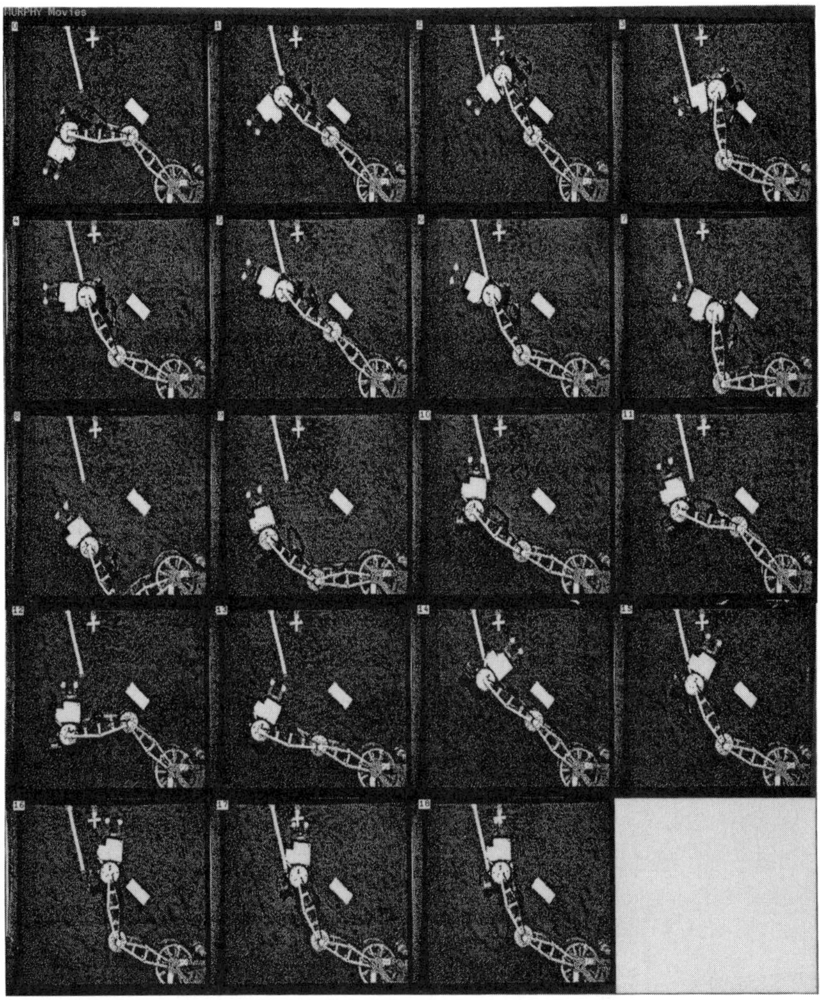

Figure 4.9: MURPHY's reach for an obstructed visual target. Slight rearrangement of the obstacles from fig. 4.8 makes that solution impossible, and forces a major detour in joint space in order to find a workable solution.

toward what appears to be a promising path over the top of the vertical obstacle (frames 8-12). As is the nature of heuristic search, the final solution paths shown here are not optimal (e.g., as short as possible), but are generally of reasonable length and directedness. No smoothing or off-line optimization of paths was implemented. Unfortunately, the level of "intelligence" necessary to understand *how* the solution paths are suboptimal is beyond MURPHY's current ability to reason. Figures 4.5 through 4.9 illustrate MURPHY's behavior on a variety of other reaching problems.

Discussion

The search space for this type of reaching problem is large, even with only three degrees of freedom: the branching factor in the search tree is 26 (i.e., all possible steps obtained by perturbing one or more of the joints in combination, in both positive and negative directions), and the typical solution paths are between 10 and 50 steps in length. The total number of paths of length 10 from a given starting configuration is nearly 10^{14}, and of length 50, there are 10^{70}. Most are not solutions and do not contain solutions. MURPHY's searches typically entail from only 10 to the low 100's of internal search steps, each step consisting of a single run through the (virtually) parallel kinematic model and evaluation functions. This successful search performance is attributable principally to the fact that the search space for reaching with obstacles is not unfriendly, unless specifically contrived to be so (as in a maze). Thus, MURPHY's crude visual heuristics, while trivial to compute from his full-visual-field representation, define an evaluation function over the joint space that has only a modest number of local minima; only a modest degree of backtracking is therefore needed on the path to a goal. The existence of a relatively smooth evaluation function is also that which allows MURPHY to search his configuration space at a relatively coarse grain of quantization (i.e., 20°/joint-step), which reduces total path length during search without impairing his ability to find solutions. On the other hand, coarse-quantization of the search space will prevent MURPHY from finding a path for his arm when he must "squirm" through too tight a space—in these cases, he "bounces" back and forth between obstacles in his mental search, never finding the clear path "up the middle". Figure 4.3 shows typical minimum obstacle clearances below which MURPHY has difficulty maneuvering. Ideally, the quantization of search could be determined adaptively during search, as some heuristic function of the "tightness of squeeze".

Interestingly, one of the most perplexing problems in the study of biological motor control has been to determine in what type of coordinates movement plans are generated in the brain. Teleological concerns argue strongly for planning in workspace coordinates, since the characteristics of a movement are most naturally specified in terms of the layout of objects in the workspace. Unfortunately, plans made in hand/workspace coordinates must eventually be translated into joint-actuator commands, which is typically assumed to require an inverse-kinematic transformation. On the other hand, whereas planning directly in joint coordinates makes an inverse kinematic transform unnecessary, the plans themselves must become more complex in order that their effects in workspace remain well-behaved (e.g., involving straight-line hand trajectories—see [75] p. 632 for discussion). MURPHY's style of planning suggests a possible resolution to this dilemma, in which plans can be made in both coordinates simultaneously (see also [116]. MURPHY generates plans as sequences of arm configurations in joint coordinates, but evaluates them on-line as sequences of visual images encoding the state of the workspace. The availability of a fast visual-kinematic model therefore allows MURPHY to incorporate workspace constraints, such as the locations of obstacles, in the construction of joint-space plans.

Visual-Feedback Control

An undesirable feature of the reaching strategy discussed above is that there is no mechanism for guessing at *good* moves that are likely to bring the hand closer to the target. It is clearly desirable that a reaching system learn how to move its joints in order to push the hand *directly* towards an object in the visual field without having to flail aimlessly, particularly if there are no obstacles to complicate the path of the arm. As discussed in chapter 2, MURPHY is endowed with two additional populations of units that allow him to learn this inverse differential kinematic relation. Each time MURPHY moves his arm through the visual field, whether physically or mentally, he generates a new training example for this differential relation. This mapping takes as input the direction the hand moved through the visual field on the trial (coded by the hand-velocity population) combined with the current global configuration of the arm (coded by the joint-angle population), and produces as output angular perturbations to the three joints (coded by the joint-velocity population). While randomly flailing for targets, therefore, MURPHY can incrementally improve his ability to move his hand in a straight line to a visual target. Figure 4.10 demonstrates this process. Frame A shows the arm poised in a "home" position, and three practice targets. Frames B, C, and D show MURPHY reaching for each of the three targets using the simple random "hill-climbing" strategy. The following three pairs of frames E, F, and G show MURPHY reaching for three nearby targets, but now using what was learned during the three practice trials to guide the joint movements at each step. At each step in this visual-feedback control mode[1], MURPHY loads his hand-velocity population with the visual-field direction traveling from his hand to the target, and then executes the partially-learned inverse differential mapping. If a readable state appears on the joint-velocity population as a result (i.e., a state of sufficient vector magnitude), then this state is used to generate the next move. The resulting perturbation to each joint is given by the center-of-mass of the pattern of activation in each of the three joint-velocity subpopulations. Once the move has been made, MURPHY repeats the process. If a state of sufficient magnitude does *not* appear on the joint-velocity population, then MURPHY reinvokes his simple programmed control mode, and begins flailing randomly until a good move is found. In both frame-pairs E and F, MURPHY can be seen to flail briefly at one point in the trajectory, a point at which the visually-controlled inverse model was not forthcoming.

It may appear from fig. 4.10 that by incrementally learning the inverse differential kinematics of the arm, it is possible to eliminate the need for search in the process of reaching for a visual target. Unfortunately, at least two factors complicate the picture. First,

[1]The term *feedback* is used here to suggest that control information is propagated back from workspace coordinates into joint-space coordinates.

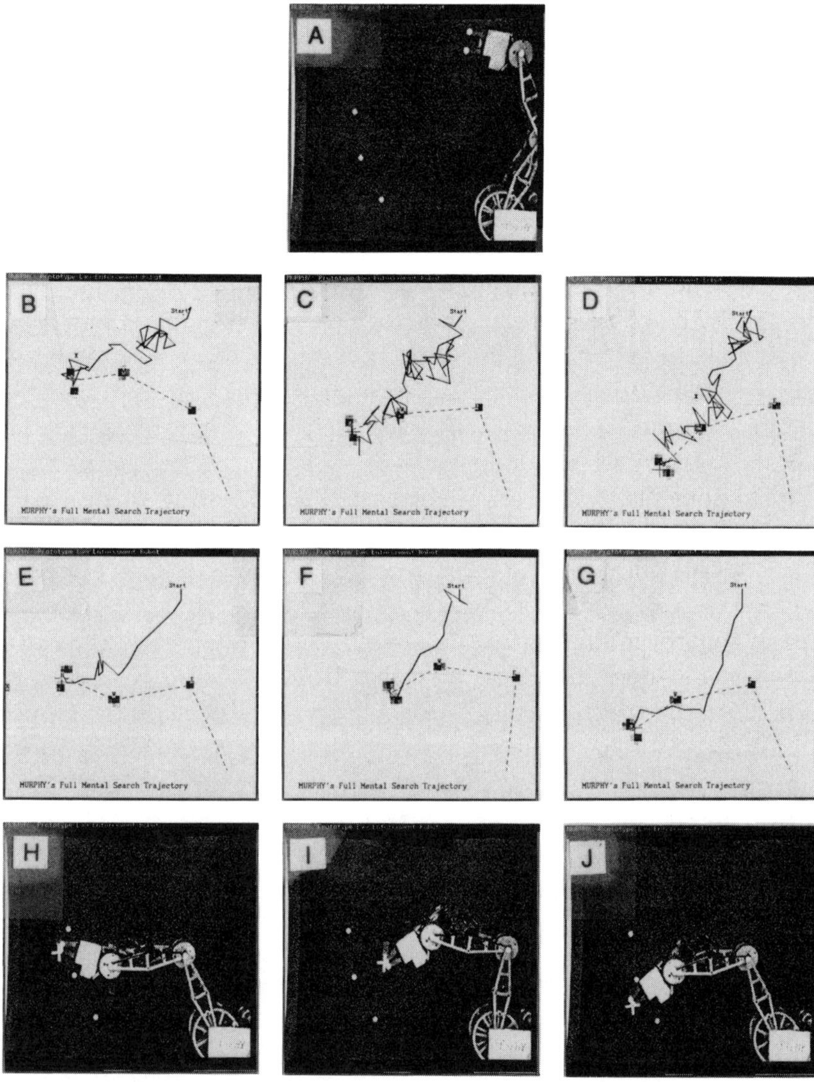

Figure 4.10: Improving with Practice. Frame A shows MURPHY contemplating three practice targets in his visual field. Frames B, C, and D show MURPHY flailing randomly toward each of the targets, but using each movement of his hand to train his inverse differential map (i.e., relating a desired vector of hand motion to the joint commands capable of producing it). After these three practice runs, MURPHY reaches for three nearby targets (frame-pairs E, F, and G), using the inverse mapping at each step to generate joint commands. Generalization to new movements is good since the underlying inverse functions vary smoothly across neighboring configurations of the arm.

the inverse differential map is in general different for every different global configuration of the arm. For example, when the arm is tangled up in a complicated way, the joint commands necessary to move the hand vertically upward may be unrelated to those needed to move the hand vertically upward with the arm unwound and starting from "home". As such, the inverse learning problem has a significantly larger input space (5-d—3 for the arm configuration plus 2 for the visual-field direction) than the static forward-kinematic model of the arm (only 3-d), therefore potentially requiring a much larger number of training examples. For the case of a 7-*dof* arm moving through a 3-d workspace, the inverse differential relation would become a 10-dimensional learning problem. The practical significance of this fact may be somewhat less severe than this theoretical argument would imply, however. Real-world reaching experiences are not uniformly distributed throughout the space of all possible arm configurations, but are rather heavily lumped in relatively small submanifolds of the joint space. For example, vastly more reaching experiences begin with the hand comfortably by the side and project to the space in front of the body—in comparison to reaching experiences where the arm begins in an arbitrarily tangled initial position and projects to yet another arbitrarily tangled final position. While it may be too expensive to learn a full 10-d inverse relation then, it may yet be possible to thoroughly learn the relation throughout the frequently visited regions of configuration space that account for a great proportion of stereotyped reaching behavior. MURPHY's learn-it-where-you-need-it approach, exemplified as well in [143], allows limited storage capacity to be conserved for the most useful regions of configuration space.

The second complication in the visual-feedback control scheme is the assumption that it is clear in which direction the hand (and other joints) should be moved in the context of a given reaching problem. While this assumption is valid in many simple cases, such as in the unobstructed trials of figure 4.10, it is in general not sufficient to move the hand directly towards a target; in may in fact be necessary to move the hand directly *away* from a target should the layout of obstacles call for it. Furthermore, the specification of an appropriate movement of the arm may require the cooperation of parts of the arm other than the hand, such as the wrist, elbow, and shoulder. MURPHY's inverse map specifies only the motion of the hand; in order to be of maximum utility, it would therefore be necessary to i) learn the inverse differential map for a set of additional control points, such as the other joints and/or links, and ii) incorporate a global critic capable of suggesting directions of movement for each of these additional points. The precomputation of global potential fields for each of several control points has been proposed by Barraquand et al. [16] as one technique for guiding a multi-link arm among obstacles to a goal configuration. However, this approach requires the predetermination of the final coordinates of each of the control points in order to establish a corresponding potential well. This precondition is somewhat paradoxical in the sense that the goal locations of control points in general should depend strongly on the unknown path to the goal.

System Performance and Scaling Behavior

Implementation and Performance Notes

MURPHY runs on a SUN 3-160 color workstation with 16MB of memory. The complete software system consists of 15,000 lines of *C* code, which includes a variety of peripheral utility routines. In k-d tree mode, a single pass through the forward visual-kinematic map, i.e., the time necessary to generate a single mental image of the arm given the joint angles, is between one and two seconds. With obstacles, the heuristic visual analysis and cosmetic color graphics processing takes up to three additional seconds, primarily reflecting the inefficiency of the algorithms used. A reaching episode with obstacles that has been contrived to be as difficult as possible may take up to 20 minutes to complete. When no obstacles are present, MURPHY typically requires less than a minute to find the target. It is estimated that the execution could be speeded up by a factor of between 10 and 50 on the existing machine by eliminating obvious gross inefficiencies. On a Connection Machine or other parallel hardware, MURPHY could easily be made to run faster than real time. MURPHY has not been optimized fora high precision of hand positioning, but even under the worst of circumstances (i.e., 8 weeks after training with numerous jostlings of the camera, tables, arm, etc.), MURPHY's hand would typically miss the target by no more than 1 inch at 7 feet (0.68° of visual angle), and would usually come within 0.5 inches. The absolute accuracy of hand positioning is a complex function of a number of system variables, and is touched upon below in the analysis of scaling of learning times.

Scaling MURPHY Up

While MURPHY's performance in a complex 2-d workspace is well within the range of current real-time processing capabilities, it is crucial to estimate how well the algorithms and data structures will scale to more realistic reaching problems. We have therefore analyzed the quantitative scaling behavior of all relevant aspects of MURPHY's architecture and performance, including size of internal representations, learning time, amount of stored data, and search time for visually-guided reaching problems. For reference, MURPHY currently uses a total of 4,517 units (4096+24+373+24), learns the 3-*dof* forward-kinematic map in 5 hours, digests 800,000 integers in the representation of the forward-kinematic functions (610 per visual unit), and solves reaching problems in his 2-d workspace in from 10 to the low 100's of internal search steps. We consider here how these numbers will scale to a 6-*dof* arm capable of reaching in a true 3-d workspace under the control of a binocular vision system.

Internal Representations. The principal visual representation will need to encode binocular visual information in order to guide the arm in depth; following the design of the primate visual system, this will involve creating a number of disparity planes, each of which contains units coarsely-tuned for a particular value of retinal disparity (for a introduction to binocular visual processing see [164]). Evidence from the biology suggests three broad classes of disparity-tuned neurons (near, far, and fixation); we liberally estimate the need for ten such classes at most. Since the units are coarsely-tuned, this implies a large number of discriminable disparity values (i.e., many more than 10; see [12, 80] for review of the coding properties of coarsely-tuned populations). It may also be desirable to employ visual units sensitive to spatial scales more appropriate to that of the arm than has been used in the current implementation. Ten disparity planes implies a 10-fold increase in the number of visual-field units to 40,960, and direction-selective visual units to 240. Both joint populations would simply be doubled in size for the jump from 3 to 6 degrees of freedom—resulting in a total of 794 joint-angle and joint-velocity units. However, if the additional degrees of freedom were to add significant length to the arm, a higher resolution would be needed in the joint populations assigned to proximal joints in order to maintain adequate resolution in the workspace. We therefore double again our estimate of joint units to nearly 1600. Thus MURPHY's total unit would be expected to increase to approximately 43,000.

Learning Time. MURPHY's learning time is dominated by the forward visual-kinematic map. The time spent learning this map over the entire joint space is proportional to d^k, where d is the arm's degrees of freedom and k is the number of distinct angular positions visited for each degree of freedom during training. In future implementations, it would seem reasonable to adopt a strategy whereby the kinematic map is learned only where it is needed, rather than for all joint configurations at once at the outset, regardless of their utility (or probability). In either case, what determines the necessary training density k? This question is impossible to answer in general; the training density needed to accurately learn MURPHY's visual-kinematic functions varies across the joint space, and depends on the intricate interplay of several rather complex, system-specific parameters.

First is the smoothness of the camera-driven visual activation functions associated with each visual-field unit (hereafter referred to as simply a unit's activation function), as plotted over the space of joint variables (fig. 3.7). Training must be more dense (and the width and spacing of joint-angle receptive fields must be correspondingly shrunk) in any region of the joint space where *any* of the 4,096 activation functions has large second derivative. This measure is needed in order that the fine structure of the activation functions be sampled with sufficient density where they are most sensitive to the joint angles. Alternatively, as was done in the current implementation, training step size (and joint-angle receptive fields) may be kept uniformly dense at a high level, as determined by the worst-case requirements over the entire joint space. This simple strategy is, however, wasteful of units and training time. It must also be pointed out that the smoothness of the visual-kinematic functions is not a given, absolute quantity, since it depends entirely on the chosen visual representation of the arm, which in turn depends on the general spatial scales of the arm's physical features, and on the response properties assigned to the internal visual-field neurons. Visual activation functions become smoother when the visual neurons are tuned to features with relatively large spatial scales, since, for a given joint-angle step size during training, large visual features move relatively more slowly through large visual receptive fields than do small visual features through small receptive fields. Why not, then, grow the visual receptive fields as large as possible, thereby lowering the needed density of training examples? In a system such as MURPHY with only a single visual representation, receptive fields can grow too large, such that visual resolution is lost where it is needed for feature analysis, such as for distinguishing arm from target from obstacles, etc. Biological visual systems, in contrast, have dozens of visual representations [206], some with small, others with very large, receptive fields. This division of visual labor is an eminently sensible engineering response to the current dilemma, and would allow a future implementation of MURPHY to significantly reduce the needed density of training examples.

A second design parameter that interacts with the smoothness vs. training-density question is the inherent approximation power of the *sigma-pi* unit, k-d tree, or other function approximator used to learn the visual-kinematic activation function at each of the 4,096 visual-field locations. MURPHY's runtime implementation uses a k-d tree for the activation function at each visual-field location (i.e., pixel). For a given activation function, the k-d tree partitions the joint space into regions, where within each region, a constant average function value is stored. This yields a piecewise-constant approximation for each of the visual-kinematic activation functions. For this type of approximation, the functions must be sampled (i.e., trained) at whatever density in joint-space is necessary in order that, in between neighboring joint-angle steps (which we would wish to be large), the visual-kinematic functions can be well approximated as constants. Steps in joint space must therefore be very small in regions where these functions are changing rapidly. It is also possible, however, to use linear or higher-order approximating surfaces locally within the regions of joint-space to fit the visual-kinematic functions. The move to these more powerful approximatorsFunction Approximation can greatly reduce the density of needed training examples, but at the expense of increased computation time during both learning and performance phases of operation (and a more clear-cut departure from neural plausibility, should this be a concern).

To make things more complicated still, it is not possible to specify in advance some required level of mean-squared error in the approximations to MURPHY's visual-kinematic activation functions. "Sufficient" accuracy in these approximations can be specified only operationally, i.e., by that which leads to internal visual images of sufficient fidelity that they may be successfully used for motion planning. This operational definition would be extremely difficult to describe analytically for a real-world system of any complexity. The level of needed accuracy in the approximations of the *individual* visual-field activation functions is much lower than might be expected, since the interpretation of the resulting image of the arm, for purposes of planning, is not based on a single, possibly inaccurate, pixel value (computed by a single visual-field unit), but rather depends on a considerable degree of spatial averaging of the values of a number of neighboring pixels for such operations as specifying the location of the arm.

In sum, a number factors can influence the training time of a given learning system such as MURPHY in complex ways. Given the difficulty of anticipating the relative importance of these various design issues in advance for a completely redesigned and much more complex system, we reason conservatively as follows. We note that the range of movement of a typical primate's arm is more limited than MURPHY's (i.e., an average of 150°/*dof* vs. 230°/*dof*), and on this basis assume the number of training positions per degree of freedom could at least be reduced from around 25 to around 10. (The reader can get a feel for this density of training examples by moving one of their joints through its full range of extension in 10 steps, remembering that is is only necessary to generalize across images from immediately neighboring joint configurations. We assume that more astute choices for the spatial scales of the visual-field unit response properties could be made to allow for adequate generalization among neighboring arm-configurations even at this reduced sampling density.) This particular choice implies one million training steps, which would take around 2 weeks instead of 5 hours. Real time on a serial computer is not, however, a particularly relevant unit of measure for MURPHY's learning process since it is entirely unit-parallel, and is therefore amenable to an immediate speedup of at least 4096. Still, for purposes of comparison, visually-guided reaching in human infants takes on the order of 4 months to develop [211, 204].

Stored Data. The total amount of data generated for the forward kinematics is equal to the total number of training steps, times the average number of activated visual units per training step, times the number of stored bits per function value (i.e., activation level). For MURPHY, theses numbers are respectively, 17,000, 50, and 8—in k-d tree form, total data consumes 7 Mbytes. Going to more realistic 3-d vision and a 6-*dof* arm will increase the average number of active visual units during training. First, the arm may be made more completely visible in order to eliminate the explicit post-computation of link volumes in the internal image. Naively, this could increase storage by a factor of 10. On the other hand, the use of a much smaller number of visual units with larger receptive fields better tuned to the spatial scale of the arm (as discussed above) could result in a large savings in both units and stored data, perhaps also by a factor of 10. We do however expect a modest increase in the number of active units due to the expansion of the internal image of the arm into several active disparity planes. For our estimate, we assume activity in three adjacent disparity planes, giving an increase of a factor of three in the number of active visual units per state. The number of training examples will grow by a factor of 60 (from above). Total data will therefore increase by a factor of 180, for a grand total of 150 million table entries. Simple extrapolation from MURPHY's storage requirements would suggest a need for 1.25 Gbytes (i.e., 2-3 CD-Roms), though actual data storage requirements should be significantly reduced by using k-d trees with least-squares linear approximators in leaf nodes. Counting *sigma-pi* synapses instead of bytes, we arrive at the figure of 500 million synapses for the 6-*dof*, binocular system. Should this estimate seem astronomically high, or unbiological, is is useful to note that there are more than 10^{14}, or over 100,000 billion synapses in the brain.

Search Time. It is difficult to estimate the search time for reaching problems in a 3-d workspace with a 6-*dof* arm. The advantage of excess degrees of freedom is that it becomes easier to "squirm" a path to a target while avoiding obstacles. Put another way, there are more directions in configuration space that take the arm "downhill" in the distance-to-target measure. On the negative side, there are more local minima, i.e., ways the arm can get tangled up in such a way as to be prevented from moving the hand incrementally closer to the target. In addition, much more complex, maze-like reaching problems could in principle be contrived in three workspace dimensions, though there is no *a priori* reason to believe real 3-d problems *are* in general more complicated. Andrews & Hogan [6] suggest on the basis of empirical observation that the advantages of additional degrees of freedom with which to squirm out of "tight spots" win out over the tendency for more elaborate ways to get tangled up in local minima. The marked success of simple heuristic search in the planning of paths for a 31-*dof* manipulator [16] is also strongly supportive of this possibility.

With the exception of the search phase, then, MURPHY's scaling behavior is relatively straightforward. Estimates for the space and time requirements of a 6-*dof*/3-d vision MURPHY system fall within the resources of state-of-the-art computing systems. The penalty for scaling to this more realistic case is primarily felt in the time necessary to cycle through the space of arm configurations during training, as well as in the increased volume of training data. It is clear, then, that MURPHY's algorithms are not suited for systems with 100 underlying degrees of freedom. In this case, learning becomes in any case intractable in the absence of an *a priori* kinematic model, since there is simply not time to gather enough training data over a 100-dimensional space. As we will see below however, traditional methods in robotics encounter severe problems of their own under these drastically increased task requirements in respect to both kinematics and collision-free path planning.

Chapter 5

Robotics Issues

In this section we review the ways in which MURPHY's algorithms and data structures relate to previous work in robot kinematics and path planning.

Learning vs. Built-in Models

The bulk of research in robotics since the birth of the field has been devoted to rigorous mathematical analysis of the kinematics and dynamics of multi-jointed manipulators, and to the development of analytical schemes for their control [39, 159]. In essence, a classical control scheme entails that the complete structure of the mechanism be solved in advance using laws of geometry and physics, and expressed in the language of explicit mathematics. Where complete advance knowledge of a manipulator is available, and in particular when the manipulator itself can be designed to advance specification, the classical equation-based approach to robot control is particularly relevant. When the structure of the manipulator cannot be known in advance, is difficult to model analytically, or changes over time, the ability to *learn* the kinematic and dynamic behavior of the manipulator becomes of critical importance—the issue of learning is therefore key in distinguishing among schemes for robot control. MURPHY is one of a growing number of systems that learn manipulator behavior by commanding the arm to move in initially unknown ways and observing the results [1, 9, 170, 47, 98, 115, 135, 143, 199, 166, 180].

Among those systems dealing primarily with arm kinematics, MURPHY is further distinguished by his emphasis on the use of a *forward* kinematic model as the primary tool for motion planning. The difficulties associated with the problem of inverse kinematics are thus avoided.

A potential disadvantage of learning is that computation time and storage requirements can grow rapidly with the manipulator's degrees of freedom, exponentially in the worst case for lookup-table methods such as MURPHY's (however, see chapter 3, page 39). Without some prior knowledge as to the form of the functions to be learned, it becomes crucial that the input-output relations are locally smooth and/or relatively sparsely-valued in the input space. In short, unbiased learning methods generally work only when the functions to be learned can be well approximated by storing a reasonable number of representative data points, with local smoothing providing generalization to novel inputs in the neighborhood of stored examplars. This is typically possible only in problems of relatively low input dimension. Where the dimension is higher, learning can still be useful if knowledge of the structure of the input-output relations is available, especially if the learning problem is linear in the parameters [9]. However, in a true high-dimensional nonlinear control problem, a built-in model becomes essential since there is simply no time or space to learn from examples.

Style of Representation

Using the Full Visual Channel

In comparing the current state of the art in robotics to the sensory-motor control systems in animals, one difference seems particularly striking. The difference is in the emphasis placed on *sensory* processes, particularly vision. Traditional methods in robotics have typically reduced the sensory representation of the workspace to the simple position and orientation of the manipulator tip, or to a small set of control points. Moreover, once this "sensory" state has been encoded, such as to specify the desired final location of the manipulator, arm control computations often proceed blindly, i.e., without further reference to the sensory input channel. Motion-planning algorithms capable of moving a multi-link arm in the presence of obstacles have typically finessed the sensory problem, assuming advance knowledge of the shapes, locations, and trajectories of all obstacles in the environment. It seems odd that the role of vision for robot control should be limited to such simple (or unexplained) functions, since it is unlikely that it will be possible to achieve true sophistication in robot control in real environments (i.e., with moving targets and partially occluded, non-polyhedral objects) without dealing squarely with sensory processing and integration issues.

The undervaluation of sensory processes in robotics seems particularly anomalous when we look at the primate brain. First, the visual input channel to the primate brain is approximately ten times "wider" than the output interface to the musculature. In addition, visual processing is by far the dominant modality in terms of sheer neural real estate, although this is partly due to the fact that vision plays many different roles in behavior other than the control of movement. The fact that vision may be a more important and difficult problem in the control of movement than the control of movement itself, may follow from the fact that the motor plant (i.e., the body) is relatively well identified and unchanging through time, while the environment on which the body must act changes from moment to moment with the position and orientation of the body, head, and eyes, and the layout of potentially moving targets and/or obstacles. This view further illustrates the need for complete, seamless integration of the sensory and motor modalities for intelligent control.

Style of sensory and motor representation is therefore a second dimension along which to distinguish work in robot control. In stark contrast to the analytic tradition, as well as to other recent connectionist schemes [1, 9, 47, 98, 115, 135, 143, 199, 166, 170, 180], MURPHY's architecture places overriding emphasis on the use of the *full visual channel*, in which a rich spectrum of information relevant to typical visual goals is either explicit or easily computed. In essentially this vein, the recent work of Barraquand et al. [16] uses a workspace bitmap for collision detection and for the construction of potential fields for obstacle avoidance. While the workspace representation of the arm used in this work is generated using a traditional analytic kinematic model, the spirit of the approach is similar to that discussed here.

As previously argued, the image-like character of a visual representation can make it ideal for capturing the rich array of heuristic visual information useful in a wide range of common motor control tasks. For example, essentially all of the information needed to direct the movements of the arm in a cluttered environment is available in relatively explicit form in an image-like representation of the visual field. This would include the positions of the hand, arm, and body relative to the objects in the environment, as well as the patterns of optic flow that specify the movements of the arm and body in the visual field and the movements and relative movements of other objects. From such representations, progress of the arm towards a visual target, as well as events such as target interception and collisions with obstacles, can be easily computed.

In addition to their utility for visual computation, at least four other advantages derive from the inherent properties of highly decoded connectionist representations. These include, i) simplified synaptic learning (e.g., [135, 115, 143, 149]), ii) fault tolerance with respect to isolated synaptic or unit failures, iii) the sufficiency of components with relatively poor dynamic range (see Hinton [80]), and when organized as topographic maps, iv) ease of local feature-extraction, such as edge detection. *However*, it must always be kept in mind that in the absence of massively parallel hardware, all of these representational advantages may be outweighed by the correspondingly massive increase in storage requirements and processing times encountered when running on a serial computer.

Two Other Connectionist Approaches

Two other connectionist schemes have addressed the learning of robot arm kinematics [115, 170] that bear certain similarities to MURPHY. Reeke et al. [170] describe a simulated automaton (Darwin III) based on an elaborate system of interconnected neuronal populations, including rough analogues for motor cortex, two subgroups of cerebellar cortex, kinesthetic and tactile representations, analogues for visual cortex and superior colliculus, and various drive-reinforcement populations. The emphasis in this work has been learning based on Darwinian selection acting on initially unstructured patterns of neural connections. The task for which this approach appeared most successful was in the learning of the inverse differential kinematic relation mapping a visual-field vector into perturbations to the joint angles, allowing the 2-joint robot to acquire visual targets. In these early experiments, the authors do not, however, address the standard problem of multiple inverse solutions for a redundant arm, nor do they study the problem of planning in the presence of obstacles.

Kuperstein [115] also describes a connectionist architecture that learns the inverse kinematics of a robot arm. This system is based on highly-decoded populations of sensory and motor units, including binocular eye-muscle and direction of gaze maps, a binocular retina with orientation-selective responses, and a set of joint-muscle units. The 5-*dof* robot arm strikes a series of random poses in view of a binocular vision system, a cylindrical target object in hand. The inverse kinematic map is learned, taking the 3-d spatial coordinates and orientation of the target cylinder into corresponding joint angles for each of the 5 joints. The visual target location is encoded by fixating both cameras on the target, and neurally representing the eye-muscle tensions and the angle of gaze. The visual orientation of the target is computed by simple oriented filters computed over the binocular visual images. After 3-5,000 training trials the inverse kinematic map is learned, such that, when the target is placed in the visual field in some orientation, a set of joint angles that send the hand to the target are directly computed by the internal connectionist map. The resulting system contains a large number of weights, 300,000 for the visual map alone.

As in [170], the standard problem of multiple inverse solutions is not addressed in this work. That is, no mechanism exists for choosing among the many configurations of the arm that can solve a given reaching problem. Consequently, the excess degrees of freedom of the arm are unused. In addition, the architecture does not appear to be extensible to the case of reaching in the presence of obstacles. It is useful to note for purposes of comparison that, while the style of representation in this work is superficially similar to that used by MURPHY, the functional role of the parallelism in Kuperstein's visual representations is primarily one of simplifying synaptic learning through a high degree of input-space decoding (see chapter 2, page 14), rather than of facilitating the computation of image-based visual heuristics.

Styles of Robot Motion Planning

In order to place MURPHY's simple heuristic-based reaching algorithm in perspective, it is useful to compare to methods that have been previously used to attack the problem of robot motion planning in the presence of obstacles. One approach is to move the arm directly towards a target, detect collisions or impending collisions with obstacles using proximity sensors on the arm, and then alter the arm's trajectory according to some pre-specified rule. Lumelsky [128] has proposed such a scheme, proving convergence for planar 2-*dof* arms of varying kinematic configurations. While this type of scheme is appealingly simple from an algorithmic standpoint, it lacks a crucial feature of *planning*, i.e., the ability to carry out advance mental operations that improve the chance of first-time success in the execution of an action or sequence of actions. Schemes of this type are essentially blindfolded maze-running algorithms, presupposing no advance or global knowledge of the environment. Motion-planning algorithms that *do* carry out advance planning computations fall generally into two types, local and global. Each is discussed below.

Artificial Potential Fields: Local Methods

One technique that has been used by several workers for simple path planning with obstacles is the use of artificial potential fields [6, 154, 104]. In this scheme, a robot is represented as a point mass. Attractive potential "wells" (representing targets) and repulsive potential "hills" (representing obstacles) are used to govern the robot's motion through space. The potential field approach is particularly well suited to the case where a global motion plan has been precomputed, such that the robot need only swerve from its intended path to avoid unexpected stationary or moving obstacles that impede its progress or threaten collision. Within this approach though, it is possible for the robot to become caught in local potential minima, for example, trapped in the elbow of a wedge-shaped obstacle that impedes further direct progress to the goal. This is therefore inherently a "local" motion-planning scheme, insufficient by itself to capture global path information. On the other hand, the necessary computations are mostly parallelizable and can be made very fast—the potential field approach can therefore be thought of as an important real-time complement to an intelligent global path planner.

In spite of the limitations of this straightforward potential field approach, the concept of explicit attractive and repulsive forces controlling the actions of a manipulator is an intuitively satisfying one, closely paralleling the natural "high-level language" of obstacle avoidance. In a sense, MURPHY's remaining-distance-to-goal search heuristics are analogous to the fields established in this class of potential field model. Unlike the potential field models, however, MURPHY's state transitions are not governed by "physical" laws embodied in an energy landscape, but by *logical* laws—in short, an algorithm. Where the algorithm prescribes downhill steps, MURPHY moves closer to his goal; where the algorithm detects an obstacle, MURPHY attempts to swerve around it; where the algorithm prescribes "uphill" steps, such as in untangling his arm caught in a local minimum, MURPHY moves uncomplainingly away from his goal, ignorant and indifferent to any laws of "physics" that might have been violated.

Geometric Motion Planning: Global Methods

Representing Free-Space

According to its standard definition, the problem of global motion planning for a multi-jointed arm in a cluttered workspace has proven to be an exceedingly complex and expensive proposition. Motion-planning algorithms almost universally proceed by building a *free-space* graph that captures the global topological structure of the problem at hand, as determined by the form of the robot, and the shapes, positions, and trajectories of the workspace obstacles. The graph essentially characterizes the set of candidate paths that the robot may use to reach the goal from its starting point while avoiding obstacles. Once built, the graph is searched for an optimal path.

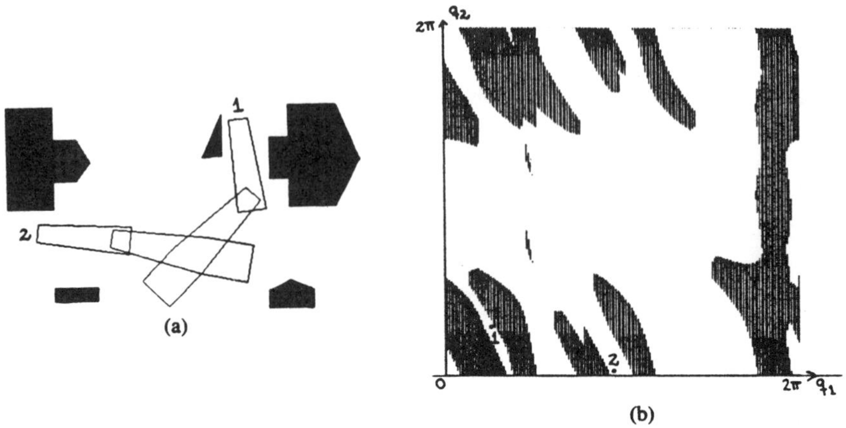

Figure 5.1: (a) Two-link revolute manipulator and obstacles. (b) Corresponding two-dimensional configuration space with obstacles (dark areas) representing illegal joint configurations. From *A Simple Motion-Planning Algorithm for General Robot Manipulators*, by T. Lozano-Pérez, *IEEE Journal of Robotics and Automation, RA-3*, pp. 224-238. ©1987 IEEE.

The most difficult and time-consuming operation is generally the construction of the free-space graph.

When the robot is assumed to be a point or a simple shape with no internal degrees of freedom, the procedure for building such a graph is greatly simplified. Brooks [29] analyzes the case of a 2-d polygonal "robot" moving amongst polygonal obstacles in a 2-d workspace. First, generalized cone primitives is used to represent the shapes of the free space between obstacles, by performing pairwise comparisons among the stored list of workspace obstacles. For each cone, the range of allowable orientations of the robot are computed (i.e., orientations of the robot such that it fits completely inside the cone as it moves down the entire length of the cone's central spine). A graph is then built, linking together intersecting cones. Finally, a path from the start to goal locations is computed by searching this graph with the A^* algorithm, allowing transitions from cone to cone only where the ranges of legal robot orientations overlap.

Motion Planning for Multi-Link Arms

When the moving robot is assumed to be a multi-link arm rather than a simple, fixed polyhedral shape, the computation of candidate paths through free space becomes much more involved. Only a handful of practical motion-planning algorithms have been devised for multi-link arms based on explicit geometric modeling of the arm and workspace obstacles [30, 51, 126, 127]. In the geometric approach, a multi-link robot arm is represented as a set of linked polyhedra, and the configuration of the arm can be represented as a point in a k-dimensional configuration space representing the arm's k degrees of freedom. Much of the machinery of the geometric approach is dedicated to determining in advance which configurations of the arm are legal, and which cause collisions with obstacles in the workspace. These "bad" arm configurations can be represented as regions in configuration space called *configuration space obstacles* (fig. 5.1). Path planning for the arm is thereby transformed into the problem of finding clear trajectories in configuration space for the arm, now represented as a point, that link the initial and final configurations of the arm while skirting configuration space obstacles. Algorithms capable of explicitly computing configuration space obstacles for a multi-link arm are in general very complex.

For example, the algorithm of Lozano-Pérez [126] proceeds as follows: 1) Assume both the initial and final goal configurations of the arm are known in advance for a particular reaching problem (presupposing an inverse kinematic solution for the arm's goal configuration is available). Also assume the positions and shapes of all objects in the environment are known; obstacles are assumed to be composed of convex polyhedra. 2) Determine the regions of the arm's configuration space that are illegal, i.e., that represent arm configurations that cause collisions with obstacles. To accomplish this, step through all configurations of all but the most distal joint at some level of quantization (e.g., 2°), and for each, compute the legal linear range of the last, most distal joint, if any. For each quantized global arm configuration, the arm is "grown" in a simple approximate fashion in such as way as to guarantee that its new, larger polyhedral volume includes the curvy "swept volume" that would be generated if all the joints were to sweep simultaneously through their small quantized ranges of orientations around the given starting configuration. If this new synthetic object does not intersect any of the workspace obstacles, then the arm is "safe" in that quantized range of configurations. If any part of the grown arm does intersect a workspace obstacle, then the configuration is marked as illegal. Once the configuration space obstacles have been computed in this way, a graph is generated that captures the topological structure of the cluttered configuration space. This graph is then searched with A^* to find the minimum distance path.

While this type of geometric approach is an algorithmic tour de force, its applicability is limited to relatively static, well-modeled environments, where several special assumptions are made that violate our ecologically-motivated design policy. These are, i) complete foreknowledge of the forward and inverse kinematic structure of the arm and the shapes, sizes, and locations of all obstacles, ii) workspace objects and arm parts consisting of convex polyhedra, and iii) a general-purpose computer for the implementation of the various geometrical computations. In contrast to the type of environment that might be found in a factory, where this type of geometric algorithm is of greatest relevance, reaching for targets in the unstructured "real world" is a very different sort of problem. For example, since the control of reaching in animals is a body-centered operation, the layout of obstacles effectively changes with every movement of the body. Objects in the world are also not polyhedra in general, and not always in full view, conditions which do not seem to pose serious problems for biological visually-guided reaching faculties. Moreover, accidental physical contact with obstacles during reaching is generally less problematic for humans and animals than for industrial robots, and optimality of obstacle-free paths is rarely a high priority in comparison to convenience of execution or quickness of response. In short, it seems likely that highly competent vision-based path planning for a multi-link arm is achievable without explicitly pre-computing illegal configurations of the arm through an involved procedure such as that described above, a procedure that grows exponentially more expensive with increasing degrees of freedom, and that must be repeated from moment to moment to reflect any changes in the relative layout of the body, target, and obstacles. In an important sense, this general approach to motion planning is inefficient: the configuration space obstacles constructed in Lozano-Pérez' geometric approach make up a global data structure that captures geometric constraints due to obstacles over the arm's *entire* configuration space, even though the arm will typically visit only a minute portion of the space during an actual reaching trial. The worst-case asymptotic complexity of this processing step for an arm with k degrees of freedom, in which the joint ranges are divided into r values, in which the robot description has m faces and edges, and in which the workspace has n faces and edges, is $O(r^{k-1}(mn)^2)$. In the case where the environment (or the location of the arm relative to the environment) is changing through time, the cost of recomputing this elaborate data structure moment by moment could easily become prohibitive. Also, exponential slowdown with increasing degrees of freedom k means that the full-blown configuration-space structure cannot practically be computed for arms with more than two degrees of freedom. Lozano-Pérez suggests heuristics for pruning the traversal of configuration space that help to reduce the computational requirements of his algorithm, but whose pragmatic range of applicability in cluttered environments is difficult to judge from the published report.

Hybrid Methods: Incorporating Heuristic Search

The principal cost of the geometric approach is the combinatorics of computing both legal (and illegal) arm configurations in advance of planning, such that the search for a path may be carried out on a much smaller graph that contains only legal arm configurations. The graph is kept as small as possible, since in order to find the optimal path, it may, in the worst case, be necessary to traverse the graph in its entirety. The algorithms and data structures needed to carry out such a program are extremely intricate, and require the direct input of an engineer in the form of forward and inverse kinematic models of the arm, and other forms of explicit geometric knowledge. The local potential field methods discussed above, while much simpler algorithmically, provide no natural mechanism for extracting the robot from local minima in the motion plans. A hybrid approach to motion planning involves the use of potential fields computed in workspace coordinates for their heuristic value, such as to suggest *likely* paths for the arm. This information can then be used in conjunction with an explicit search procedure, including the global geometric techniques discussed above [16, 51, 93, 210].

Hwang & Ahuja [93] describe one such approach to motion planning. A scalar potential function similar to those described above is constructed from the layout of the target and obstacles in the workspace. The minimum potential valleys running between obstacles are located in the resulting field; these minima capture the topological structure of the free space. Each such potential valley represents a candidate path for a moving robot. A graph is then built and searched using dynamic programming to determine a best first-guess at a complete path from start to goal. Various heuristics and optimization techniques are then applied to generate collision-free trajectories that minimize length and change of orientation. The advantage of this approach is that the computation of initial guesses as to candidate paths is relatively inexpensive in comparison to purely geometric approaches, which generally perform large numbers of explicit geometric computations involving the shapes and intersections of polygons (or polyhedra in higher dimensions).

Another interesting approach to motion planning similarly based on potential fields and heuristic search, and that shares certain representational features in common with MURPHY, is that of Barraquand et al. [16]. In this work, multi-resolution bitmaps are used to represent both the workspace and the configuration space (fig. 5.2). Global potential fields are constructed for each of a number of control points on the manipulator, providing the heuristic "pull" on each such point towards its goal location and away from obstacles. An arbitration scheme is used to settle conflicts among the various potential field gradients, and a randomized search procedure is used to extract the system from local minima when they occur. The authors demonstrate successful motion planning for simulated manipulators with up to 31 degrees of freedom, a truly impressive result.

The approach of Barraquand et al. [16] is similar to the current work in the sense that it is based on direct search through the arm's configuration space, guided by simple geometric heuristics computed

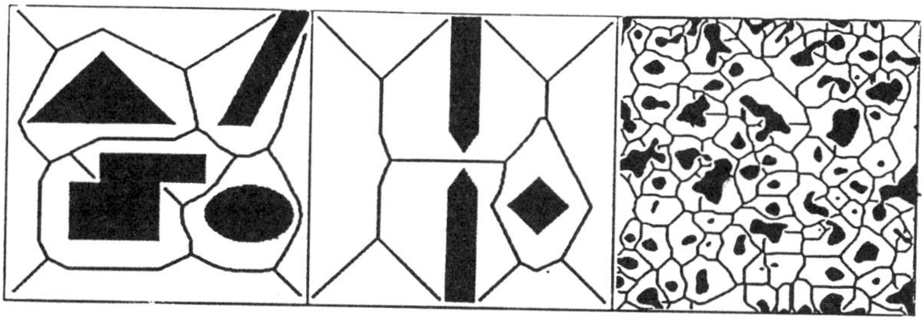

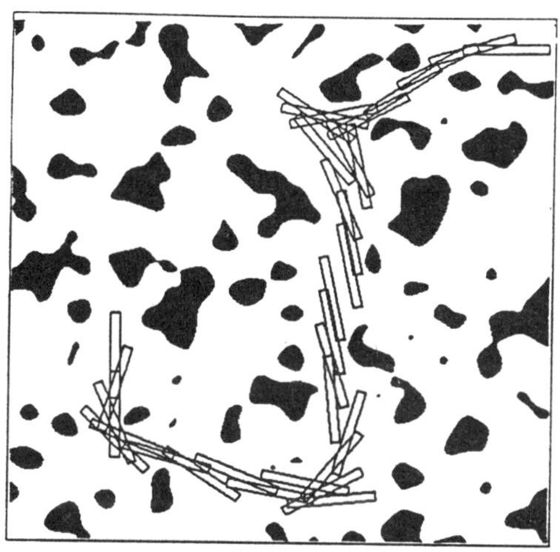

Figure 5.2: Using numerical potential fields for robot motion planning [16]. In this approach, a potential field is constructed over a workspace bitmap for each of a set of control points on the robot. Each field has only one global minimum at the goal location. Each potential field is then skeletonized (3 frames at top) such that gradients to the goal steer clear of obstacles. The potential fields for each of the control points are then combined, giving rise to a composite potential field that is used to guide the search. Lower frame shows a complete path found for a mobile robot. Reprinted with permission. ©J. Barraquand, B. Langlois, & J.C. Latombe, 1989.

over an image-like representation of the workspace. However, the two approaches also differ in several ways. For example, the work of Barraquand et al. does not address the problem of kinematic learning, instead using built-in forward and inverse kinematic models for their simulated manipulators of various descriptions. Furthermore, the inverse kinematic model is crucial in the work of Barraquand et al., since it is used to compute goal configurations for the arm that determine the potential "attractor basins" for each of the control points. In contrast, MURPHY searches forward from the starting configuration with no specific goal configuration for the arm in mind (except the constraint that the hand must acquire the target). MURPHY's distance-to-goal evaluation function, while similar in function to a potential field, is never fully and explicitly computed over the entire workspace, but is instead evaluated only for those states that the arm actually visits along the search path. Finally, the built-in data structures and procedures used in [16] are far more heterogeneous and elaborate than the connectionist substrate used by MURPHY, in part a reflection of very different underlying design constraints and motivations.

MURPHY Uses Direct Heuristic Search

In spite of these differences, the current approach shares with approach of Barraquand et al. certain general characteristics. In brief, both methods avoid the expensive pre-processing step associated with traditional geometric techniques that partition configuration space into forbidden regions vs. free space. Instead, a much less expensive potential field computation is carried out that provides a heuristic distance-to-goal evaluation function over the workspace [16]. In actuality, MURPHY does not explicitly precompute such a potential function over the entire workspace, but rather computes a crude measure of remaining distance to the goal at each step during planning that serves as a virtual potential field. The crucial point is that an easily computed potential field can be of good heuristic value, making it possible to search directly for paths without excessive backtracking. MURPHY, for example, is content to visit legal and illegal arm configurations alike during motion planning, i.e., is perfectly content to "bash" into an obstacle when he makes a wrong turn, since the collision is a purely internal event. This is the classic advantage of a mental model: simulating reality can be faster, cheaper, and/or safer than doing it for real, and increases the chance of success on the first actual try. In order for search to be efficient, it is only necessary that the computations involved in running the internal model and computing the heuristics be as fast as possible. Within MURPHY's architecture, both of these operations consist of straightforward parallel operations. Additionally, unlike the explicit geometric approaches to motion planning, the workspace obstacles in these approaches can be of arbitrary shape at no extra cost, since they are simply treated as blobs in the workspace that the arm is intended to avoid.

In this direct search formulation, extra work is introduced in the search process only when backtracking becomes necessary to extract the arm from a local minimum in the heuristic evaluation function. In MURPHY's case, this cost has proven to be acceptable empirically. The most difficult reaching problems that it was possible to construct in MURPHY's planar world (and that MURPHY was able to solve—see section 4) required in the low hundreds of internal steps through configuration space, giving rise to plans of between 10 and 50 steps (e.g., fig. 4.4). For purposes of comparison, we estimate using Lozano-Pérez' complexity bound of $O(r^{k-1}(mn)^2)$ that problems of this approximate difficulty would, in a standard geometric approach, consume in the tens to hundreds of *millions* of serial machine operations. This estimate is derived under the following assumptions: the constant of proportionality in the complexity bound from [126] is estimated to be 6×10^{-7} seconds from the problem of fig. 5.1, which the author states consumed 6 seconds of total wall-clock time on a Symbolics 3600 for (apparent) parameter values $r = 180$, $k = 2$, $m = 8$, and $n = 30$. For a 3-*dof* (i.e., $k = 3$) problem with identical obstacle layout but with MURPHY's much coarser grain of quantization (i.e., $r = 18$), we estimate a wall-clock time of 11 seconds for this algorithm, or, in the tens of millions of machine instructions. However in making this comparison, we must carefully consider that i) much in the serial/geometric algorithmic approach could in principle be parallelized, and ii) the emphasis on path optimality is in great part that which adds computational complexity to the serial/geometric approach.

Interestingly, in contrast to such geometric motion-planning algorithms, which as indicated above grow exponentially slower with increasing degrees of freedom in the manipulator, direct heuristic search methods may become *easier*. As degrees of freedom are added to an arm without adding joints (for example, the human arm which has 7 degrees of freedom distributed across the wrist, elbow, and shoulder), the arm becomes more "snakelike". In this case, more paths become possible since it becomes easier to "slither" around corners and in and out of the nooks and crannies that frustrate simpler arms. On the other hand, direct heuristic search has a generic tendency to become *more* expensive with increasing degrees of freedom, since, for a given step size (i.e., search space quantization), the branching factor for enumerating next moves increases rapidly. It is difficult to precisely predict the relative weights of these two factors.

Chapter 6

Psychological Issues

Many aspects of MURPHY's design have been inspired by the brains and behavior of animals. However, to this point we have discussed MURPHY almost exclusively in an engineering context, i.e., as a set of algorithms and data structures designed to attack a difficult computational problem in sensory-motor control. Engineering goals aside, visually-guided reaching behavior is of great ecological significance to humans and other primates, who spend a large fraction of their waking hours manipulating objects (and other primates) in their immediate extrapersonal space. Significantly, the concept of extrapersonal space has both psychological and neural reality. Piaget was perhaps the first to recognize the special psychological status of the space immediately surrounding the body, i.e., within reach, showing how the child's construction of his own reality depends on active manipulation and query of the immediately surrounding environment during the sensorimotor stage of development [160]. In the brain, neurons in the posterior parietal cortex have been identified that respond vigorously to visual stimuli only when the stimuli are of motivational significance and fall within arm's reach [151, 174] (see chapter 7, page 120). From a neuropsychological perspective, Roland et al. [176] have shown using PET scans that these same posterior parietal areas are strongly activated only during movements or imagined movements made relative to objects in surrounding space, such as moving the arm through a maze, and not for movements in general.

In this and the following chapter, we explore the degree to which MURPHY's "biologically inspired" structure and function are, after all, relevant to the mechanisms for vision-based limb control in humans and other animals. We show the natural mechanisms for intelligent limb control are in some ways remarkably similar to those built into MURPHY, much more so than anticipated by this author during MURPHY's design. Beyond these comparisons, moreover, we show that MURPHY's architecture and algorithms can help bring into relief and render meaningful several features of process and representation in the biological control of limb movements that might otherwise have seemed arbitrary, or that might have gone altogether unnoticed in the data.

Psychoanalyzing MURPHY

As was stated at the outset of this report, MURPHY's organization and behavior, while "biologically inspired", have not been intended as specific neural or psychological models for the control of limb action in a particular animal species. Indeed, MURPHY's psychology is quite *unlike* that of his natural counterparts in a number of ways: i) motor learning in humans and animals is not typically accomplished during a single uniform practice session over the space of actions, ii) where learning and performance phases of activity are distinct for MURPHY, they coexist and co-evolve in biological organisms, iii) in reaching for obstructed targets, humans and other primates clearly make more informed guesses in choosing initial candidate paths for their moving arms than does MURPHY, iv) likewise, there is no evidence that primates precompute entire complex trajectories mentally before actually moving their arms; a much more realistic scenario would have the mental model keep *just ahead* of the physical arm trajectory in order to avert collisions, and v) MURPHY lacks any interesting treatment of goals and subgoals, clearly a crucial aspect of intelligent manipulation.

In spite of these differences, important similarities exist as well that are most evident at a more abstract level of description: i) basic sensory-motor learning requires an extended period of active physical practice [72, 186], ii) improvement in certain types of performance is possible with either physical or *mental* practice [169, 186, 195], iii) analogical sequences of mental images can be used to solve difficult spatial reasoning problems [62, 111, 134, 189], iv) preprogramming of motor sequences takes longer for more difficult tasks [74, 107], and v) two modes of vision-based arm control can be distinguished, one a sensory-locked subsystem for direct visual-feedback control, the other a programmed-feedforward control subsystem for more complex movement patterns [66, 158]. In the following sections, we explore these issues in greater detail.

Development and Plasticity in Limb Control

The Necessity of Active Sensory-Motor Learning

Perhaps the most obvious feature of MURPHY's sensory-motor learning program that is shared in common with human and animal developmental mechanisms is the fact that visual control of motor behavior can only develop or adapt when, during learning, changes in visual input are systematically related to self-produced movement [21, 71, 72, 73]. Thus Bauer & Held have shown with monkeys that active experience with the arm *in view* is necessary for the development of accurate reaching [21]. Similarly, cats that are prevented from seeing their paws during early development can learn to walk normally and avoid obstacles, but cannot accurately position their paws during a visual placing response (i.e., the characteristic paw extension elicited when a cat is lowered towards a horizontal surface) [70]. In adult humans, adaptation to artificially altered visual-kinematics of an arm occurs only with active movement with the arm in view, and not in either passive-movement or no-movement conditions [72].

Why should learning be necessary for a sensory-motor function as crucial as vision-based limb control? It is interesting to consider the similar computational pressures that may have selected for learning in both natural and artificial sensory-motor control systems. Held [72] points out that the ability to learn and adapt to the changing sensory-motor relationships that result from growth of the body, aging, injury, or evolution, is far preferable to the alternative of explicit, genetic pre-specification of these relationships for all conceivable body states. In fact, it seems difficult even to argue the possibility that the nervous system could genetically anticipate (even approximately) all possible sensory-motor configurations, including all those that have been contrived by experimental psychologists. From an engineering perspective, the ability of a robotic system to learn by doing allows the designer to ignore the myriad details of the sensory-motor apparatus, details that are in general extremely difficult to model (e.g., slippage in gears, inaccuracies of manufacture, non-rigidity, friction, spherical aberrations in the imaging system), and that can change over time. Thus for both the natural and artificial cases, the ability to learn allows the same "brain" to be inserted in a wide range of "bodies".

Developmental Evidence for Two Modes of Control

The pattern of development in infants of the ability to reach for visual targets can provide interesting clues as to the nature of the underlying information processes [211], and possibly suggest similarities and differences with respect to MURPHY's learning behavior. Non-functional proto-reaching movements that are very weakly correlated with vision are present at birth in humans ([82], see [203] for review), but while this implies some degree of hardware predisposition to the task, the disposition appears to be limited to a slightly increased tendency to project the arm forward in the presence of a visual stimulus [82]. Any obvious influence of vision on movements of the arm typically disappears soon after birth, re-emerging at the end of the fourth month in the form of highly effective visually-guided reaching movements [203] (fig. 6.1). Of great interest and relevance to the current issue is the fact that while infants at this stage are highly proficient at projecting their hands to unobstructed visual targets, they are unable to reach around even the simplest barriers to acquire an object of interest. This more "intelligent" capability does not emerge until late in the first year of life, i.e., four to six months after the onset of simple reaching [125]. Furthermore, these older infants have more trouble reaching around transparent than opaque barriers (manifested by their tendency to reach straight towards the target in spite of the transparent barrier), suggesting poorly regulated competition among two distinct modes of control.

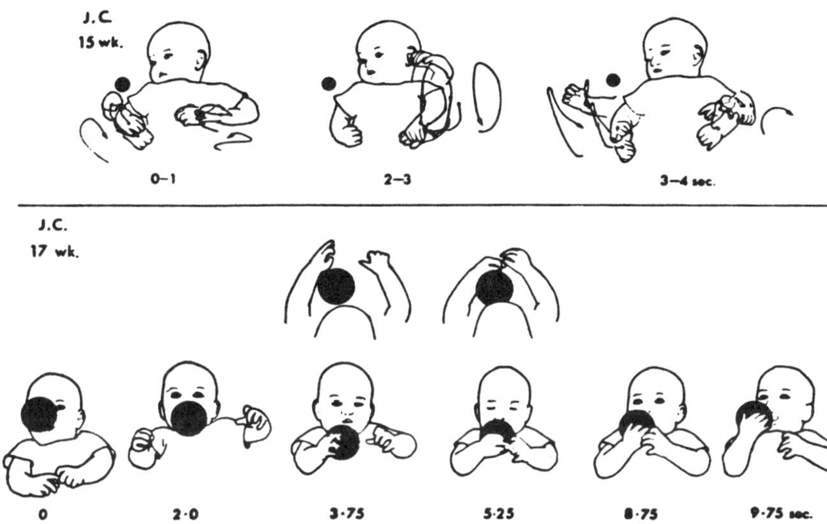

Figure 6.1: The emergence of visually-controlled reaching. Top row shows poorly controlled attempts to grasp a small ball at 15 weeks. Bottom row shows same baby encircle and pull in a suspended ball 2 weeks later. Reprinted by permission of the publisher. From *Basic Patterns of Psychogenic Change in Infancy*, by C. Trevarthen, in *Regressions in Mental Development: Basic Phenomena and Theories*, edited by T.G. Bever. ©1982 Lawrence Erlbaum Associates.

The developmental evidence in human infants thus suggests a partitioning of vision-based arm control into two modes, analogous to MURPHY's stimulus-locked visual-feedback vs. programmed-feedforward control modes. That such a functional partitioning does in fact exist in both humans and animals is supported by independent neural [66] and psychophysical [158] evidence (see section 7); the distinction has not yet, to the knowledge of this author, been explicitly made in the developmental literature. This analogy to MURPHY's principal functional subdivision is of particular theoretical interest, since the evidence for dual processing modes in biological limb control was not known to this author at the time of MURPHY's design. Far from planned in advance, MURPHY was only endowed with a visual-feedback control subsystem to compensate for obvious inefficiencies in his previously existing feedforward-only design [135].

The superposition of visual-feedback and programmed feedforward modes of control is eminently sensible from both ecological and computational standpoints, and it is therefore unsurprising that it should "co-evolve" in both natural and artificial systems for visual limb control. The visual-feedback subsystem is fast, useful for swatting, fighting, catching, and unobstructed reaching; the programmed mode of control keeps the arm from colliding with obstacles, and is necessary for movement where the immediately present sensory information cannot prescribe the appropriate action in a simple way—in short, cases that require inference or "thought". *A posteriori* it is not surprising that the visual-feedback mode of control should be the first of the two to come on-line in human infants, since the ability to internally program movement sequences is clearly the more demanding information process.

Possible Evidence for Local Learning

Why should sensory-motor learning take so long? In order for MURPHY to learn the kinematics of his relatively simple 3-*dof* arm from the perspective of his relatively simple cyclopean visual system, he must exercise his arm through tens of thousands of positions over a period of several hours. Were MURPHY scaled to 6-*dof*'s, the learning time could increase to a period of days or weeks (see chapter 4, pp. 69-74). Given that learning times of this magnitude seem uncommonly long within the field of robotics (in the few cases where learning approaches have been studied, e.g., [9]), and enormous in comparison to typical machine-learning applications, it seems clear that the MURPHY's learning scheme is surely flawed, or at minimum highly inefficient. However, in comparison to time frames observed in the development of reaching in human infants, these learning times seem, if anything, too short.

In MURPHY's case, a relatively long initial learning period is a direct consequence of the table-based learning algorithm which generalizes only locally. While the use of a lookup-table or its neural equivalent is a very general approach to function learning, data must be "laboriously" gathered from across the entire space of input variables (in MURPHY's case, the 3-d space of joint configurations). Generalization to novel inputs is only local, i.e., based on proximity to previously seen inputs. For example, MURPHY's ability to run his visual-kinematic model in one part of joint space is independent of his ability to run it in another, distantly separated part of joint space. This distinguishes this type of learning scheme from a parameter-estimation method that might attempt to fit the visual-kinematic functions globally with a small number of trainable parameters. Learning models of this latter type are characterized by rapid learning and global generalization (e.g., [9]). Human infants require four to five months of visual-kinematic experience before accurate visually-guided reaching is possible, and another four to six months before the onset of even rudimentary programmed visual control. Though it is difficult to separate the effects of biological maturation from the effects of learning, such prolonged learning times could be interpreted as evidence for an exhaustive local learning procedure, although other explanations of course remain possible. What would constitute independent evidence that visual control of arm movements is based on a joint-space lookup-table? One type of evidence would be non-uniform proficiency in vision-based control across the space of joint configurations, which, if correlated with practice, would strongly suggest a locally-generalizing model. Effects of this kind are common in all areas of human sensory-motor control, most easily seen in the domain of athletics (e.g., having a more controlled forehand than backhand). A second type of evidence for local, table-based control would come from the absence of either positive or negative transfer from sensory-motor adaptation across distantly separated kinematic domains. While this author knows of no evidence either for or against this possibility, it would seem easily testable experimentally and of great theoretical interest. One possible experiment along these lines would be to induce adaptation of static perceived arm position with a prism as in [72], and then test for the presence of adaptation in the same arm but in progressively more different kinematic configurations. A rapid falloff in the transfer of adaptation with kinematic distance would be direct evidence that sensory-motor learning is based on a locally-generalizing internal representation of the space of limb configurations.

Models, Imagery, Practice, and Stability

Building and Using Mental Models

One of this author's principal motivations in choosing vision-based robot-arm control as a domain for experimentation was the possibility of studying one of the most basic computational problems faced by behaving animals that involves what may be described as "thinking". What is thinking? We digress momentarily to consider the thoughts of a handful of eminent scientists on the subject, beginning with William James in 1890:

> "The pursuance of future ends and the choice of means for their attainment are thus the mark and criterion of the presence of mentality in a phenomenon [96]".

On the question as to why thinking is useful (or why the lack of it is dangerous), James writes, in full color:

> "The advantages of this are obvious. Take the prehension of food as an example and suppose it to be a reflex performance of the lower centers. The animal will be condemned fatally and irresistibly to snap at it whenever presented, no matter what the circumstances may be; he can no more disobey this prompting than water can refuse to boil when a fire is kindled

> under the pot. His life will again and again pay the forfeit of his gluttony. Exposure to retaliation, to other enemies, to traps, to poisons, to the dangers of repletion, must be regular parts of his existence. His lack of all thought by which to weigh the danger against the attractiveness of the bait, and of all volition to remain hungry a little while longer, is the direct measure of his lowness in the mental scale. And those fishes which, like our cunners and sculpins, are no sooner thrown back from the hook into the water than they automatically seize the hook again, would soon expiate the degradation of their intelligence by the extinction of their type, did not their exaggerated fecundity atone for their imprudence."

Craik [40] was both more reserved and more explicit in his discussions of thinking and its advantages, and first introduced the concept of a *mental model*:

> "If the organism carries a 'small-scale model' of external reality and of its own possible actions within its head, it is able to try out various alternatives, conclude which is the best of them, react to future situations before they arise, utilize the knowledge of past events in dealing with the present and future, and in every way to react in a much fuller, safer, and more competent manner to the emergencies which face it."

Piaget speculated as to the nature of a child's mental models, and explicitly acknowledged the possible utility of visual mental images (synopsis of Piaget's writings taken from [65]):

> The notable achievement of stage 6 [18 months to 2 years] is the appearance of the capacity to represent mentally an object or action which is not perceptually present... In the first place, the infant when faced with new models [i.e., things to imitate] no longer needs to perform overtly his trial attempts at imitation; instead, he now tries out the various movements mentally. When he has made the necessary mental adjustments, he can then perform the correct action. Since the process is largely mental, the stage 6 infant can imitate more quickly than the one who must first try out all the movements. The internalization of the trial and error process consequently leads to what appears to be an immediate imitation of models... That is, the infant is capable of evoking (representing) the absent model in

> some internal symbolic form, for example, by means of a visual image."

Holland et al. [85] draw a similar conclusion in more modern language:

> "In common with many recent theoretical treatments, we believe that cognitive systems construct models of the problem space that are then mentally "run", or manipulated to produce expectations about the environment... The cognitive system attempts to plan a sequence of actions that will transform the initial problematic state into a goal-satisfying state. An adequate mental model can accomplish this task by mimicking the environment up to an acceptable level of approximation."

Common to all of these accounts is the notion that intelligent behavior is centrally based on the ability to "run internal simulations" using a "small-scale model" that "mimics the environment up to an acceptable level of approximation"—possibly using "visual images"—in order to "plan a sequence of actions that will transform the initial problematic state into a goal-satisfying state." Needless to say, the foregoing language is a blueprint for MURPHY's algorithmic makeup. A particularly interesting challenge has been to combine within a single architecture the advantages of both parallel and serial computation. Inside MURPHY, a sequential controller is used to exercise a parallel mental model (fig. 2.2). Rumelhart et al. [183] also discuss the complementary roles in cognition of parallel and serial computation, but arrive at a somewhat different conclusion. In their formulation (fig. 6.2), sequences of states are generated by a parallel mental model that feeds back on itself (via an intermediary "interpretation" network).

Mental Imagery

A further challenge in MURPHY's development has been to propose novel functional roles for visual imagery in a difficult real-world robot planning problem. A simultaneous goal has been to suggest, through a concrete computational model, a novel biologically-plausible role for imagery in the vision-based control of limb movements. Mental

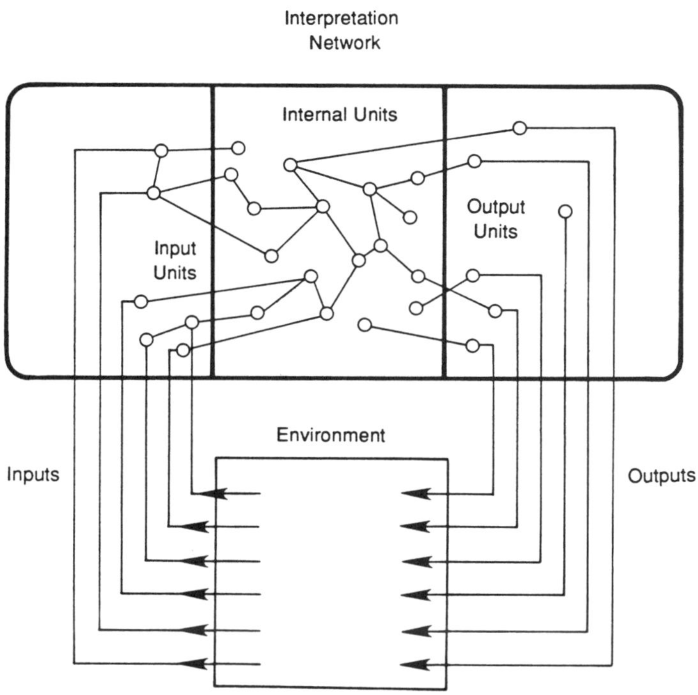

Figure 6.2: Sequential behavior can be achieved through feedback in a parallel mental model. Reprinted by permission of the publisher from *Schemata and Sequential Thought Processes in PDP Models*, by D.E. Rumelhart *et al.*, in *Parallel Distributed Processing: Explorations in the Microstructure of Cognition, Volume 2*. ©1986 MIT Press.

imagery phenomena have been extensively studied within the behavioral sciences, with emphasis on both cognitive [57, 111, 112, 162, 189, 190, 191] and neuropsychological [44, 49, 113, 177] or purely neural [62] issues. The existence of widespread topographic organization in visual cortex [206] and recent results showing localized cerebral blood-flow increases during mental imagery tasks [44, 176] invite optimism regarding the possibility for a rapprochement of the neural and psychological data. Despite the enormous attraction of a possible bridge between imagery as a cognitive process and imagery as a neural process, there has been a surprising dearth of concrete proposals as to how the brain might actually implement realistic imagery-based functions. Trehub [201] and Kosslyn & Schwartz [111] have proposed retina-like models capable of carrying out innate 2-dimensional mental image transformations. In previous work, this author [134] described a connectionist model able to perform 3-dimensional transformations of mental images, and whose repertoire of visual transformations (including 3-d rotation, zoom, and pan) was acquired through an initial period of sensory-motor learning. Much in MURPHY's design has been derived from this earlier model, including the emphasis on sensory-motor learning by doing. To the knowledge of this author, MURPHY is the most sophisticated working system thus far built that manipulates visual imagery, and the only such system to use real visual input.

In the remainder of this section we briefly review selected issues that bear on the nature of mental imagery and its underlying representations, once again seeking comparisons where possible to MURPHY's algorithms and data structures. We also present two engineering arguments that bolster the propositions, longstanding in the field, that i) visual imagery is at least in part mediated by array-like, metric visual representations, and ii) perceptual and imagery processes are likely to share the same neural hardware.

Metric Visual Imagery

The existence and specific nature of visual mental imagery has always been a central issue in the study of cognition (see Pinker [161] for an excellent review). The seminal work of Shepard & Metzler

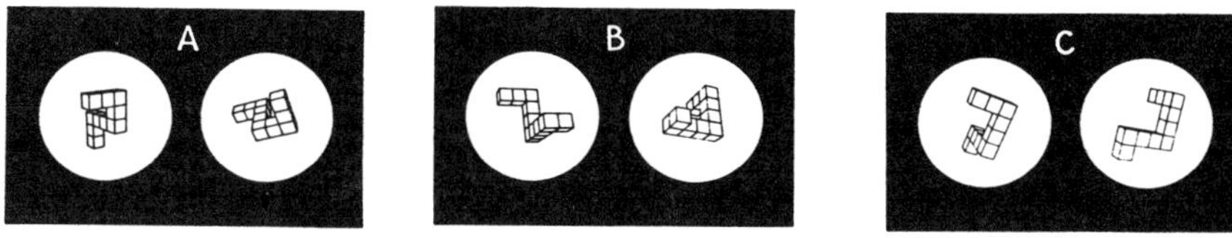

Figure 6.3: Stimuli used to study mental rotation. Pair A differs by a rotation in the image plane, pair B by a rotation in depth; objects in pair C are distinct. General result in this experiment is that the time needed to judge that two views correspond to the same 3-d object grows linearly in the angular disparity of the views, both for rotation in the image plane and in depth. Reprinted by permission of the publisher from *Mental Images and Their Transformations*, by R.N. Shepard and L.A. Cooper. ©1982 MIT Press.

[190] demonstrated that the time required to recognize that two perspective drawings that portray the same 3-dimensional object grows linearly in the angular difference of the portrayed orientations, both for rotations in the image plane as well as in depth. On the basis of this and numerous follow-up studies, Shepard and others have argued that the underlying representations used in tasks of this sort are array-like, i.e., metric (admitting of such concepts as proximity, extent, and direction), viewer-centered, analogical visual representations which, during rotation or other mental transformation, move through sequences of intermediate states that correspond to intermediate states in the world [57, 111, 112, 162, 189, 190, 191]. MURPHY's mental imagery is also array-like, metric, viewer-centered, and analogical, and also moves through sequences of intermediate states that correspond to intermediate states in the world. Further, in correspondence with the principal psychological measure used in these studies, MURPHY's mental sequences take a time that generally increases with the difficulty of the spatial reasoning problems he

faces, though measures of difficulty are much less clearly defined for MURPHY's reaching problems than for simple mental rotation tasks. Comparatively little psychological data has been gathered as to the duration of this type of advance motor-programming in humans as a function of movement complexity, but available results have generally shown that movement reaction times increase with the "complexity" of the motor act to follow, such as for arm movements [74], Morse-code sequences [106], and syllable sequences [105] (see [107] for review). It was also found that reaction times were only dependent on the complexity of the initial part of a movement sequence, implying that later parts of a movement sequence are programmed after the movement has begun (counter to an arbitrary design decision specifying that MURPHY only moves once an entire movement program is found). Unfortunately, these results, while in the anticipated direction, are only partially related to MURPHY's "reaction times" in advance of movement, since these latter times involve a problem-solving (i.e., search) component, and not just the programming of a known movement pattern. It seems that much more elaborate and definitive work on the chronometry of limb motion planning could be carried out, and this author would await such results with great interest. In a fascinating recent study, Georgopoulos and collaborators have shown through single-cell recording in M1 that the preparatory motor state of a monkey changes continuously through time in advance of reaching for a visual target whose location is derived through a 90° mental rotation of the visual field [62]. The result seems to imply that some neural area upstream from primary motor cortex, speculatively a visual area, must also be changing state through time in correspondence to the visual mental rotation required by the task.

Returning to the issue of metric visual imagery, it seems necessary on the basis of extensive experimental data to conclude that visual mental imagery is at least partly mediated by an array-like representation of the visual scene [57, 111, 112, 162, 189, 190, 191]. The possibility that the brain does *not* use this type of array-based imagery at all becomes even more unlikely in light of the following series of observations: i) the brain has clearly gone to a great deal of trouble to maintain topographic order through several levels of the visual system [206], ii) topographic visual representations are ideally suited on pure engineering grounds for a range of visuo-spatial computations, such as mental rotation, scanning, etc., that are known to be carried out in the brain (e.g., [111, 134]), iii) biological design is known to be highly conservative, exhibiting a strong tendency to recruit existing structures in the solution of new problems, and iv) the style of connectionist architecture necessary to take advantage of such visual-field representations exists in the cerebral cortex (see chapter 7). In addition, much more specific evidence is taken up in chapter 7 that suggests concrete roles for cortical visual and motor areas in the generation and use of metric visual imagery.

In contrast to the foregoing account of metric, analogical mental imagery, Hinton [76] has argued *against* the existence of an array-like substrate for visual imagery and in favor of a more abstract, structural description of the visual scene, on the basis of evidence that mental images of complex 3-dimensional objects do not necessarily contain all of the spatial relations that should in principle be present in a pictorial, array-like rendition (see also [78]). Thus, while work on mental imagery has produced one of the most compelling and extensive lines of evidence regarding the specific nature of any single type of cognitive representation, there remain important areas of disagreement in the interpretation of this evidence [161]. It seems likely that much of the remaining controversy is due to the fact that in the brain, the visual world is represented simultaneously in several different forms whose effects superpose transparently during normal visual behavior, a fact that should greatly complicate the task of the experimental psychologist. In support of this view, Kosslyn [113] and Farrah [49] have argued that far from being a unitary computational process, mental imagery is dependent on the interactions of several independent "subroutines" including, among others, image-generation and image-transformation modules. The existence of multiple, superimposed subsystems dedicated to visual function is far from unusual: for example, i) the several neural subsystems for oculomotor control (saccadic, smooth-pursuit, convergence , vestibularly controlled eye-movements—see [88] for review), ii) the "what/where" distinction that separates the roles of the occipital-parietal vs. occipital-temporal visual pathways roughly into pattern-analysis vs. spatial visual processing modes [146], and

iii) as will be seen in section 7, the considerable evidence for two parallel subsystems dedicated to distinct modes of vision-based limb control [66]. Thus, a disentanglement of the full repertoire of visual imagery functions involved in normal behavior awaits future work.

Imagery vs. Perception

Beyond the question as to structural nature of the representations that support imagery function, a related issue that has engendered significant debate is the question as to whether visual perceptual and imagery functions share the same neural hardware. On the basis of a variety of psychological data, a number authors have concluded that this is the case (e.g., [55, 50, 113]). For example, visual mental imagery has been shown to exhibit the same relation between eccentricity and resolution as holds for visual perception [56], and has also been shown to function equivalently to visual percepts in producing adaptation (see [55]). Cerebral blood-flow studies have also shown that mental imagery tasks selectively activate neural areas that are known physiologically to be driven by visual input.

A very basic feature of MURPHY's organization may lend additional insight into this question, by exemplifying a case in which the hardware overlap of perception and imagery is obligatory. MURPHY's ability to generate mental images on his visual-field population is entirely dependent on a prior learning period during which the same visual-field population was driven "unconditionally" by visual input from the camera. From the perspective of MURPHY's cortically-inspired architecture, the question as to whether visual imagery and perceptual functions share the same hardware is a definitional "yes". To the extent that the mental imagery functions of the brain are also acquired through visual-motor experience, there seems no reason to believe the same argument would not also hold for the same reason.

Mental Practice: Learning by Thinking

The ability of human subjects to significantly improve their performance relative to controls on a variety of motor skills through purely mental practice has been extensively documented in the motor learning literature (see [45, 186, 195] for reviews). Given MURPHY's ability to also improve performance in his visual-feedback mode of control through covert or "mental" practice (chapter 4), illuminating comparisons can once again be drawn. To review briefly, MURPHY first acquires the direct kinematic mapping that relates commanded limb postures to the corresponding visual images of the arm, across the entire space of postures. Once learned, MURPHY can run this "mental model" internally to train his visual-feedback mode of control, which relates the desired *direction* of change in visual hand position to the necessary commands for movement at each of the joints. What MURPHY learns here is essentially the ability to move his hand along a desired visual trajectory, a clear requirement in many visual-motor skills. Covert learning of this visual-feedback mapping is possible since knowledge of the visual effects of *changes* in posture is implicitly present in the ability to generate views of arbitrary *stable* arm postures. Thus, in internally simulating his arm moving through sequences of neighboring stable postures, MURPHY can covertly learn about the visual effects of changes in posture.

We now turn to mental practice effects in human learners. For a wide variety of motor skills, such as serving a volleyball, passing a football, shooting freethrows, and tossing rings, subjects who have been instructed to *imagine* performing the skill without overt movement can be shown to incur statistically significant (and sometimes large) positive transfer to skill in the actual task [45, 186, 195]. One of the most striking examples of this mental practice effect has been shown for a rotary-pursuit task [169]. Three groups of subjects were given 25 trials on day 1. On days 2 through 9, a physical-practice group was given 25 trials per day, a mental practice group practiced only by imagining and visualizing the task, and a no-practice group received no practice of either kind. On day 10, all subjects were tested with 25 additional trials. The principal result of this experiment is depicted in fig. 6.4. The mental practice group experienced essentially the same improvement as did the physical-practice group, and both practice groups, not surprisingly, significantly outperformed the no-practice group.

Mental practice effects are not seen for all tasks, however. It appears that prior familiarity with the task is an important variable in determining the degree to which mental practice by itself can improve

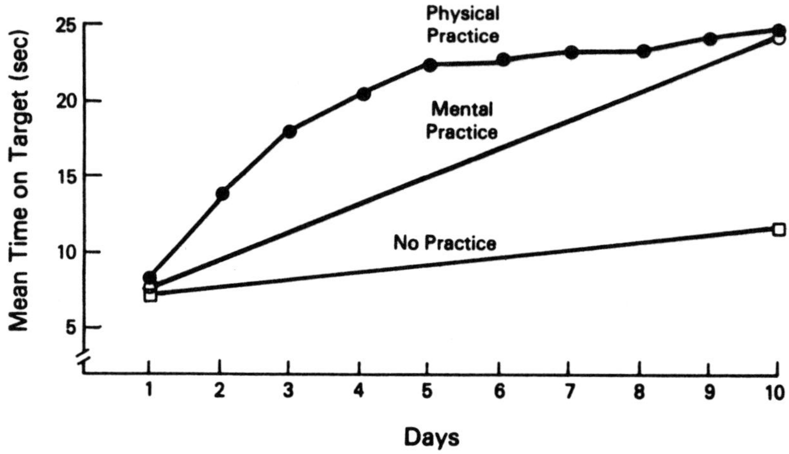

Figure 6.4: Effect of mode of practice on acquisition of a rotary-pursuit coordination task. From *Psychonomic Science, 26*, pp. 71-73. Reprinted by permission of Psychonomic Society, Inc.

performance (i.e., when *not* used in conjunction with physical practice). Thus, for unfamiliar tasks, mental practice is of little use, and has even been shown to be deleterious to performance [64]. Clark [38] showed that the relative benefits of mental practice in a basketball task increased with prior skill level; novices benefited more from physical practice, while more skilled subjects improved nearly as well with mental as with physical practice. From the perspective of mental modeling, this familiarity effect is not surprising: in order for a mental model to be useful for covert learning, it must first exist. Thus, just as MURPHY can only learn covertly after first acquiring a repertoire of basic visual-kinematic relationships, it appears that human learners, too, must gain initial visual-kinematic experience with the task at hand before they are able to "practice" the task internally.

There is also some suggestion that mental practice is more effective for tasks that involve simpler or more restricted movements, than for those in which movement dynamics come into greater play [35, 144], although the available data is not definitive. Thus, Brumbach [35] showed improvement with mental practice only for a short badminton serve, and not for a long badminton serve (requiring greater movement amplitude and speed) or for putting a golf ball (requiring more precise control of force). Minas [144] also showed mental practice to be of minimal value in a motor task that involved throwing balls of varying weights into the proper bins, a task where the control of varying loads is required. (Mental practice in this task did, however, lead to significant improvement in the execution of task rules.) These results are at minimum consistent with the conclusion that the "mental model" subjects use to internally simulate the execution of a visually-guided arm movement task does not allow for the accurate simulation of varying loading conditions or of movements of such force or amplitude that the dynamical properties of the limb come into significant play. In other words, it appears to be difficult for subjects to imagine the effects of arm and load dynamics. This could imply that i) the mental model does not have loading conditions as input parameters—if it did, performance would be uniform across loading conditions and movement speeds—but rather, is specific to a particular set of learned loading conditions (e.g., with badminton racquet in hand), and ii) the mental model is essentially visual-kinematic, encoding the relationships between joint positions and the visual state of the arm with a fixed load, and therefore allows for only crude trajectory approximation at higher movement speeds (where arm and load masses and moments of inertia begin to exert their perturbing effects). A fascinating additional result seems also to suggest that the internal model subjects use in covertly practicing or planning limb movements is primarily kinematic, and primarily positional (rather than velocity-related). This is the observation that amputees typically experience the existence of a "phantom" limb, whose position *but not movement* can be readily "felt" during the intended execution of limb activity (see McCloskey [131] for detailed arguments). These implications are of theoretical interest in the current discussion since they bear directly on the question as to whether the brain uses primarily positional/kinematic (like MURPHY) or primarily dynamic (as for industrial robots) variables in the control of limb movements. This question is taken up again in next chapter.

Corollary Discharges and Perceptual Stability

A final psychological topic that bears mention in the current discussion is the historically significant concept of *corollary discharge* or *efference copy* [131]. In brief, corollary discharges, a term coined by Sperry [197], have been defined as outgoing, command-related (e.g., motor) signals generated in the central nervous system that also directly "irradiate" higher sensory areas, in such as way as to either modulate incoming sensory information, or to produce sensations in their own right. Sensations of this type are generated purely internally, and are therefore distinct from those that result from peripheral sensory stimulation. With respect to human psychology, and with particular emphasis on limb-related effects, the concept of a corollary discharge has been invoked primarily to explain the subjective perception of body state or movement in the absence of peripheral input, such as the sensation of "phantom" limb position or movement in amputees (see [131]). The term "efference copy" introduced by von Holst & Mittelstaedt [84] refers to a specific type of corollary discharge that produces a *negative* image of the expected sensory consequences of an action in progress. When combined with the incoming sensory stream, the efference copy is intended to cancel out the expected, or "reafferent", component of the incoming sensory stream produced by the animal's own movements, leaving only the unexpected, or "exafferent", component signaling actual change or movement in the environment. Efference copy has been used primarily as an explanation for *perceptual stability*, i.e., the fact that the visual world appears subjectively stationary in spite of a nearly constant visual flow across the moving retina. In a number of biological studies, efference copy mechanisms have been shown to act to prevent powerful self-stimulation of an animal's sensory apparatus during its own movement or action, such as in the lateral line system in the goldfish which detects fine displacements of water against the body [185], the electroreceptor organs of electric fish [23], the auditory system of echolocating bats [198], and the superior colliculus in the monkey which plays roles in both the detection and initiation of eye movements [171].

The crucial computational issue for corollary discharge hypotheses is the question as to how the outgoing command, encoded in the language of the motor (or other effector) system, is to be translated into the appropriate neural "discharge" in the language of the *sensory* system where it either results in an immediate sensation, or for the case of an efference copy, cancels out the incoming sensory image. In certain cases this translation process is trivial, such as when the sensory and motor languages are similar (e.g., isomorphic somatosensory and motor maps of the body). In other than these cases, however, no general account has been given in the neural or psychological literature, to the knowledge of this author, as to how an outgoing motor state is to be converted into a specific sensory prediction that takes into account the relations between the animal's sensory and effector apparati and the intervening environment. MURPHY represents one such account, demonstrating a general technique by which outgoing motor command-states can be mapped into predictive sensory images. Though MURPHY's "corollary discharges" give rise to positive images of his arm that are useful for motion planning, the architecture and computations that could underlie *perceptual stability* via negative efference copy are essentially identical.

Summary of Psychological Issues

A number of psychological issues have proven to be of relevance to MURPHY's structure and function, and vice versa. On balance, MURPHY's "deep" psychology, with its emphasis on learning by doing and the use of mental models and imagery, is not unrelated to that of human and animal learning and performance in certain sensory-motor domains. We leave this section with a list of specific questions for whom possible answers have been suggested by aspects of MURPHY's design:

1. Why basic learning of vision-based biological limb control requires an extended period of active physical practice with the limb in view, and how this could imply a local, table-based representation of visual-kinematics.

2. How sequences of analogical visual mental images could arise in the brain as corollary discharges from outgoing motor-command centers, and how they may play a role in the vision-based planning of limb movements.

3. How developmental patterns of visually-guided reaching in infants supports the notion that the biological control of limb movements is subdivided into two distinct subsystems, one for immediate visual-feedback control, the other for more general preprogrammed movement sequences. Also, recalling the pressures that drove MURPHY's design, why this functional subdivision makes good engineering sense.

4. How improvement in certain types of performance is possible with purely *mental* practice, why such improvement is contingent upon some initial physical experience with the task domain, and why it may be more effective for tasks that are primarily kinematic in nature.

5. Why the capacity for learning and plasticity in vision-based limb control provides additional indirect evidence that imagery functions and direct perceptual input must physically converge in the brain.

6. How the concepts of mental modeling, mental imagery, mental practice, motor programming, and "thinking" appear to be intimately related in sensory-motor behavior, all involving the use of learned relationships between command states and their sensory consequences.

Chapter 7

Biological Issues

From a biological perspective, MURPHY's role is essentially that of a theoretical "pilot study", at once a testing ground for intuition and a hunting ground for mechanism and explanation. We first briefly review in the next section what is known of the biological motor-control interface, that is, how the brain may communicate with the musculature in order to effect movement. The purpose of this review is to determine whether MURPHY's use of purely kinematic control variables (i.e., positions and velocities, while ignoring forces and torques) is a serious oversimplification of the biological control problem for arm movement, or whether, to the contrary, these variables can be made to capture much of the richness of vision-based biological limb control. We then review specifics of neural organization relevant to the control of limb action in extrapersonal space, making a number of connections to MURPHY's "anatomy" and "physiology". Particularly tantalizing are a series of functional and structural parallels to the biology that were unanticipated at the time of MURPHY's design, but which may be presumed to have arisen from similar computational pressures.

The Muscle Interface

MURPHY moves his arm with pure kinematic position and velocity control. Thus, he simply sends angular position commands to the stepper motors in each of his three joints, which drive their respective joints to the prescribed angle and then stop. Since forces and torques are completely ignored, it could be argued that MURPHY's motor interface is highly oversimplified and unrealistic in relation to the biological motor interface, particularly given the emphasis in the field of robotics on the explicit control of joint torques.

A priori, kinematic variables (i.e., positions, velocities, and accelerations) would seem to be the ideal choice for visually-guided biological limb control, particularly for the arms, since they constitute perhaps the simplest and most natural language with which to describe hand and arm motion for typical, ecologically-relevant tasks [75]. Thus, for reaching out for a piece of food, bringing food to the mouth, grooming another primate, moving objects from place to place in front of the body, or for catching a ball, the control of hand and arm position and velocity seems the appropriate level of description. This conclusion is supported by the observation that position, direction, and velocity are ubiquitously encoded in the visual system [206], which plays a crucial role in the guidance of movement. In addition, evidence reviewed in preceding sections suggests that internal planning of limb action may in fact be carried out using primarily kinematic/postural control variables (see pp. 108, 112).

The appropriateness of kinematic control for typical ecologically-relevant tasks does not necessarily hold for the control of industrial robots, that may as a matter of course need to perform complex, accurate, high-speed motions with unknown loads whose inertial parameters could significantly alter the dynamics of the arm. Standard robotics methodology explicitly incorporates forces and torques in the control process for this reason, and the joint actuators for most industrially useful robots are at the lowest level driven with variable torque commands. Thus, once a kinematic path description has been generated (i.e., as a desired time sequence of joint angles), a second, *inverse dynamics* control stage computes the required torques at each of the joints necessary to achieve the kinematically-specified trajectory [39]. There has been much debate as to exactly how these joint torques are best computed in man-made robotic systems, in addition to the question as to how they may be generated in biological motor control systems (see [9, 75, 87, 99, 186] for discussion of these and related issues).

Feedback control is conceptually simplest among control schemes. In this approach, torques are computed for each joint in such a way as to reduce the difference between the desired and actual joint positions (or velocities, etc.). An attractive feature of this type of control is that it is simple and reliable, can compensate for unpredictable disturbances, and can be relegated to a simple "reflex loop". The idea that such a reflex loop for limb control could be mediated through sensory feedback to the central nervous system has been highly influential in the field of motor physiology (see [186] for review). However, recent work has demonstrated unequivocably that complex arm movements in humans and primates can be executed more quickly than could be explained by sensory feedback ascending to the central nervous system (CNS), suggesting that the CNS can emit motor commands in a purely feedforward manner [26]. This does not, however, rule out an important role for *peripheral* feedback loops, such as sensory feedback from muscle spindles to targets in the spinal cord.

From the perspective of industrial robotics, feedback control is an important and useful technique, but is also a "weak" method in that it incorporates no foreknowledge as to the appropriate actuator commands for the desired movement trajectory. For this reason, it may be too slow for the industrially significant case of accurate, high-speed motions with substantial loads. *Feedforward* control schemes are usually used to compensate for these shortcomings [39]. These schemes generally involve an analytic model for the dynamics of the arm based on the kinematic and inertial parameters of the arm's links and their loads (i.e., link lengths, relative positions and orientations, masses, and moments of inertia). The model is used to compute in advance the joint torques necessary to move the loaded arm through its kinematically-specified paces. The analytic approach to arm dynamics is of indubitable significance for the precise control of industrial robots, but viewed from a biological perspective, raises troubling questions as to how the central nervous system is to carry out these additional complex transformations in its real-time motor control activities. In the remainder of this section, we review evidence from work in the physiology of motor control in humans and animals that suggests the peripheral nervous system is cleverly organized in order to greatly reduce the need for central feedforward control of limb dynamics. Kinematic control, while an oversimplification, may be closer to the control language of the central nervous system than the complexities of robot arm dynamics and human motor physiology might lead one initially to believe.

The Mass-Spring Model. The realization that muscles may behave like springs [8, 86] has given rise to the *mass-spring* model for movement control [26, 36, 54, 79, 83, 87](for review of concepts and physiological evidence, see [186]). In this model, the flexor and extensor muscles about a joint can be viewed as a pair of opposing springs whose equilibrium state determines the position of the joint. Movement is controlled by varying this equilibrium position over time, achieved by manipulating the relative tensions of the opposing springs. One advantage of this type of muscle interface is that arm (limb) trajectories are robust both with respect to errors in the starting position as well as unexpected mid-course perturbations, since the springs automatically generate restoring forces that attempt to keep the arm on its specified *virtual trajectory.* Most significant from the control perspective, the springs themselves are responsible for the generation of the appropriate joint torques through time, rather than the controller, which is only responsible for specifying sequences of virtual positions. Since it is mediated by springs, the equilibrium control scheme consists simultaneously of position *and* static force control, since a spring is essentially a relationship between force and length [79, 83]. Thus, when the arm is unloaded (or lightly loaded) and at modest velocities, the model can approximate pure kinematic position control. When the arm pushes isometrically against an obstructing surface, the control variable can be used to specify static force, in the direction the arm *would have moved* if unobstructed. Intermediate loading conditions can introduce trajectory irregularities as the limbs spring more sluggishly to their equilibrium positions, but result only in constant errors in final limb position. One may imagine a tennis stroke with a brick strapped to one's wrist—the stroke is likely to be misshapen, but the follow-through will simply be too low. For both isometric and intermediate loading conditions, the ability to visually control the direction of desired limb movement can be used to advantage: in the former case to deliver specific forces to obstructing surfaces, in the latter case to compensate for positional errors.

Summary. If, as the above evidence suggests, the biological motor interface can be approximated by a mass-spring control system, then MURPHY's purely kinematic feedforward mode of control becomes a valid abstraction of the biology for lightly loaded movements at reasonable speeds (accounting for the great majority of primate arm movements for which a reasonable degree of positional accuracy is required). Furthermore, MURPHY possesses two strategies for dealing with the kind of trajectory and positional errors that would be incurred by a mass-spring controller under more severe loading conditions, including i) the use of a visual-feedback controller capable of generating appropriate compensatory arm movements (e.g. to prop up an arm weighed down by a heavy load), and ii) relearning the kinematic mapping from scratch under new loading conditions, for loads that are handled very often (such as a tennis racquet by an expert player). In addition to a role in load compensation, MURPHY's visual-feedback controller could in principle be used to generate directed forces for compliant motion, i.e., motion in contact with surfaces. Suffice it to say that the combination of feedforward and visual-feedback kinematic control modes, while inadequate for heavy-duty industrial applications, appears to support many of the everyday arm control needs of primates.

A Cortical Basis for Visual Limb Control

MURPHY's connectionist architecture has been designed to emulate the gross systems-level organization of the cerebral cortex. Several neurally-inspired guidelines were observed: i) create a set of neural populations, each of which represents some aspect of the organism's sensory, motor, or internal state, ii) define the semantics (i.e., response properties) of individual neurons in these populations in keeping with known electrophysiological response properties of cortical neurons, and iii) allow for plastic, bidirectional interconnection pathways between any two (or more) of the cortical populations. It was envisioned that within such a framework, an organism could simply *behave*, and while doing so, profit from a rich set of associative

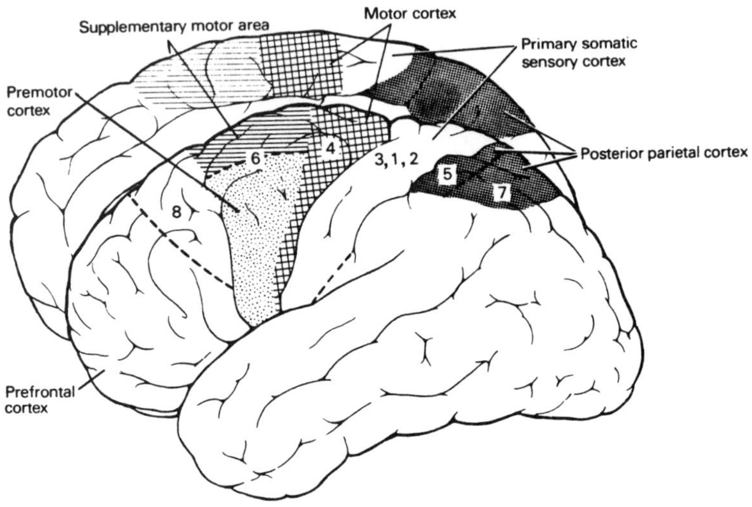

Figure 7.1: The cortical regions responsible for the control of limb movements lie near the primary motor cortex (Brodmann's area 4), including the supplementary and premotor cortices (area 6), and the superior (area 5) and inferior (area 7) lobules of the posterior parietal association cortex. Reprinted by permission of the publisher from *Voluntary Movement*, by C. Guez, in *Principles of Neural Science*, edited by E.R. Kandel and J.H. Schwartz. ©1985 by Elsevier Science Publishing Company, Inc.

learning opportunities. Once learned, these associative mappings could act as *mental models* that could be run internally to guide future intelligent behavior and thought. This very general machine architecture is essentially a clearing house for learn-as-you-go associative mappings, whose power and versatility grow with the representational diversity of the underlying neural areas, the degree of area-to-area interconnectivity, and the storage capacity of the plastic interconnection pathways.

On the basis of a number of independent lines of evidence, it seems clear that in primates, a large portion of the control of visually-guided arm movements can be accounted for in a relatively well-defined complex of cortical areas [66, 92]. These areas include area 6 of Brodmann (see fig. 7.1), comprising the premotor (PMA) and supplementary motor (SMA) areas located just anterior to the primary precentral motor area (M1) [66, 92, 175, 176, 214], and both areas 5 and 7 of the posterior parietal association cortex [92, 121, 129, 151, 176]. Unfortunately for present purposes, this complex of cortical areas is by no means entirely dedicated to the control of visually-guided reaching; the SMA and PMA have been implicated in numerous other integrative sensory and motor functions, such as the control of eye movements and the production of speech, while the posterior parietal regions also play critical roles in the direction of attention and spatial perception and memory (see [66, 129] for excellent reviews). The array of behavioral deficits resulting from damage to the posterior parietal lobes is particularly daunting, making it inappropriate to attempt to treat these areas comprehensively within a single-function computational model. Our scope in the following discussion is therefore highly selective, necessarily limited to data specifically relevant to the control of visually-guided reaching. Arbitrary selectivity of neural data is dangerous, however the availability of a working computational model for the control of visually-guided reaching, based on a neurally-inspired system architecture and built from neurally-inspired components, *even if wrong as a biological model*, affords an extra degree of discrimination in our attempts to interpret the complex and often self-contradictory morass of anatomical, physiological, and neuropathological data relevant to the issue.

One salient feature of the cortical substrate for the control of reaching is that it appears to be split into two distinct subsystems implementing two entirely different modes of control [66], also supported by developmental evidence from human infants as reviewed above. In a tantalizing analogy to the distinction between MURPHY's programmed-feedforward and visual-feedback control modes, Goldberg [66] distinguishes the "medial" SMA-based neural subsystem from the "lateral" PMA-based system:

> "The medial [SMA] system operates in "projectional" action or action that is driven forward by prediction derived from an internal model of the world composed from previous experience which permits the creation of a probabilistic model of the future [25]. The lateral [PMA] system is part of a system responsible for recognizing and associating motivational significance with external objects and, in relation to action, operates in a responsive mode in which each action is dependent upon an explicit external input."

A Programmed-Feedforward Neural Subsystem

In the following treatment of the literature, we focus first on the brain's medial, "model-based" motion-planning subsystem for limb control. To lend structure to our exploration, we review briefly the key functional components of MURPHY's predictive, motion-planning subsystem. The principal components are a representation of static joint configuration, projecting synaptically to a topographically organized map of the visual field. In addition, during all types of behavior including learning, overt motor performance, and covert motion planning, an external sequential controller drives the static joint representation through sequences of states. During learning, synapses are modified from the joint to visual populations (fig. 2.2). During motion planning, the internally driven sequences of hypothetical joint postures give rise, via the learned synaptic pathway, to the sequences of visual mental images that are used to create and evaluate motion plans.

The Supplementary Motor Area. Turning to the biology, the SMA appears to be a site for the programming of sequences of motor states involved in the preparation of complex motor acts, and may play a role similar to that of the sequential planning control structure which drives MURPHY's joints through sequences of postures [66, 175, 176, 214]. In keeping with this notion, high levels of electrical stimulation to the SMA typically elicit stable postures, commonly elevation of the contralateral arm with abduction and external rotation of the shoulder, in contrast to the phasic and rapidly habituating muscular responses evoked with stimulation of the primary motor cortex (M1) [92]. Roland et al. [175] reported that significant increases in regional cerebral blood flow (rCBF) are seen in both the SMA and in M1 during the actual execution of complex motor tasks, but that when subjects were told to *mentally rehearse* these same motor sequences, blood flow was seen to significantly increase in the SMA alone. Suggestive of its role in motion planning, extensive premotor ablation including the SMA (but also extending to the PMA), impaired the ability of monkeys to adapt a trained, visually directed straight-line reaching movement to reach *around* a transparent barrier for a piece of food [147], a task that requires nontrivial trajectory planning. Likewise, damage to the SMA or other aspect of the medial control system in humans (e.g. through Parkinson's disease) has been found to produce a more severe impairment to intentional, pre-planned actions than to actions that are in response to immediate environmental contingencies, such as visual or other sensory cues [118].

Area 5 of the Superior Parietal Lobule. Having posited a crude analogy between the SMA and MURPHY's sequential planning element, we next seek evidence for a static body image representation upon which the SMA could act, and which in turn projects directly to a topographic visual representation. One promising candidate is area 5 in the posterior parietal cortex, which has a significant reciprocal connection with the SMA. Mountcastle et al. [151] reported that two thirds of neurons in area 5 responded to joint rotation [151]. Responses were much brisker to active rather than passive movement of the joints (probably due to reafferent activation by the SMA [151]), and 82% of them were related to single joints. Many were found to be "sensitive indicators of the steady joint position, at least over the time scale of a few seconds". However, the exact nature of cortical responses to joint position are difficult to surmise from the published literature. A question of theoretical interest here is whether a population of joint units exists with peaked responses traversing the range of angles for each joint. Early reports seemed to indicate cortical joint representations of this kind; according to Rose & Mountcastle [178], units in area 5 resembled those found in primary somatosensory cortex (S1) from which area 5 receives a major corticocortical input, whose broadly overlapping receptive fields are bell-shaped in joint angle, with widths of $15 - 30^o$ [178]. However, more recent physiological studies in somatosensory cortex and in area 5 of the posterior parietal cortex indicate that the responses of posture-related neurons grow monotonically with increasing (or decreasing) joint angle, rather than exhibiting "double-ended", or peaked, response curves (for review, see [60]).

Interestingly, 5% of area 5 cells were termed "arm-projection neurons" by Mountcastle et al., firing only before and during reaching movements toward objects of motivational interest within immediate extrapersonal space, and not for arbitrary, non-goal-directed arm motion. Roland et al. [176] have also demonstrated in rCBF studies that area 5 is only activated during voluntary movements in extrapersonal (as opposed to *intra*personal) space, i.e., for movements that require that the body and limbs be "lined up" with objects in the external environmental. Finally, electrical stimulation in area 5, as for the SMA, gives rise to movements that are unlike the phasic, "twitch-like" single-muscle movements evoked by stimulation to the primary motor cortex, but rather are more slowly adapting, coordinated orienting or limb projection or withdrawal movements that may outlast the period of stimulation[1] (see [92]). On the basis of the foregoing data, and in analogy with MURPHY's joint-angle pop-

[1] It is important that the evocation of movement *per se* in these studies is not incompatible with the idea that area 5 neurons primarily encode limb *position* rather than velocity or direction of motion—movement of this type, i.e., the assumption of stable postures, is in fact prescribed by the "mass-spring" theory of motor control discussed above. Within this theory, opponent muscle groups at a joint are treated as springs whose relative tensions specify the joint's equilibrium position. When a new equilibrium position is specified, the joints will in fact move, but the induced movement is due to the difference between the current position and the specified equilibrium position, rather than due to an explicit command to move in a specific direction or velocity.

ulation, it may be concluded (see [151]) that i) area 5 is driven in part by the SMA, an area implicated in the planning and initiation of complex motor sequences, and ii) at least part of the function of area 5 is to represent body and limb posture information in the formation of movement plans relative to external, environmentally-specified coordinates.

Area 7 and the Inferior Parietal Lobule. A major output projection of area 5 terminates in adjacent area 7 and other areas in the inferior parietal lobule (IPL) [152, 5]. The IPL is not a unitary neural area, but is composed of numerous distinct subentities, including areas 7a and 7b, MT, MST, DP, and LIP, whose boundaries and functions in many cases remain obscure [5]. While area 7 is clearly implicated in a variety of integrative sensory-motor functions, including the control of visually-guided reaching, but perhaps even more prominently the control of eye movements and other attentional processes, it seems safe to say that the specificity of currently available anatomical and physiological data will at best support only vague conjecture as to the specific interactions among the cortical areas in this region in the execution of any single integrative function. Still, with our goal to complete here the rough analogy between MURPHY's predictive arm-control subsystem and that of the cerebral cortex, we note that area 7 is a major recipient of extrastriate visual information, and given its close relationship to area 5, can therefore be viewed as a candidate for the putative visual imagery functions underlying limb motion planning.

The studies of Roland & Friberg [177] in fact strongly implicate areas 5 and 7 of the posterior parietal lobe in this genre of mental imagery function, having shown increases in rCBF in these areas only during a mental visual route-finding task, in contrast to an arithmetic task or a jingle recitation task. Only the route-finding task required the envisionment of action in an environmental coordinate frame. Farrah [50] reviews further evidence that the posterior parietal lobe may be a locus of visual mental imagery, citing selective deficits in mental imagery tasks in patients with occipito-parietal lobe damage. As alluded to above, Roland et al. [176] also demonstrated that whereas the posterior parietal areas showed no significant increase in rCBF during the execution of complex sequences of finger movements (consecutively touching the fingers to the thumb), exemplifying a motor task that does not require reasoning relative to an environmental reference frame, a significant increase in rCBF *was* seen during execution of motor tasks involving coordinated arm movements in an extrapersonal reference frame (such as moving the hand through a maze). It seems difficult to dispute, then, that the posterior parietal cortex, including both areas 5 and 7, is active selectively during actions or imagined actions carried out relative to externally specified visual coordinate systems.

In keeping with the notion that the posterior parietal cortex plays a central role in the coordination of motor and sensory information for the control of movement in extrapersonal space [129, 151], it has been long been known that gross errors in reaching for objects occur, both under visual guidance and in darkness, in human patients with damage to the posterior parietal cortex, and in monkeys with experimental lesions to these areas (see [129] for review—it is to be stressed that a large array of other deficits result from this type of brain damage as well). At the same time, the ability to perform complex motor sequences with the hands and arms is not disrupted *per se*, as evidenced by the ability of monkeys to solve previously learned latch-box problems following posterior parietal lesion [43]. Once again, it appears that areas 5 and 7 together are selectively important for tasks that require the explicit relation of the body's internal coordinate frame with that of the external environment, and not for those that simply involve complex activity of the hands and arms.

Much has been learned about the visual and command functions of area 7 through single-unit recordings in awake, behaving animals. The most detailed single-unit data has been related to the control of eye-movements and the direction of attention, though a proportion of the units studied have had complex and interesting reach-related response properties. To generalize, area 7 neurons differ from those of more primary sensory (and motor) neurons in the dependence of their sensory responses on the preparatory or attentional state of the animal. Put another way, many area 7 neurons have "receptive fields" that are simultaneously sensory *and* intentional, or command related. Most of the visual units described in area 7 have been modulated by eye position or eye movement, as well as attentional and motivational variables (see Lynch [129, 153] for reviews). However, 34% of area 7 neurons were classified as *arm-projection* or *hand-manipulation* neurons similar to those seen in area 5 (see above) [151]. A variety of other types of integrative cells are present in area 7 as well, such as those with both visual and cutaneous receptive fields that are fired by either a tactile stimulus to, say, the forearm, or by the visual approach of the tactile stimulator to that same part of the arm [94].

We pause here to consider the "physiological" response properties that would be observed in recording from units in MURPHY's principal "visual" area. Whereas the units are nominally visual, i.e., driven by the camera, *after* training they also display complex joint-related responses. By definition, learning has taken place when each of the initially visual units learns to "recognize" (i.e., learns to be driven by) global joint configurations that introduce some part of the arm into the unit's visual receptive field. Put another way, these units encode the same region of extrapersonal space both visually, and in terms of the (possibly numerous) joint configurations that place a part of the arm in that region of space. It is intriguing that the command-related activity of a unit of this kind might in the standard parlance be described as "behavioral set", when in the current context the unit is more appropriately viewed as a single "pixel" in a predictive image of the sensory consequences of the limb action that is being contemplated. From an electrophysiologist's point of view, units of this complex visual-joint type should be relatively difficult to detect, since the optimal stimulus conditions are multimodal and obscure, particularly when the joint contribution is best (or only) evoked by *active* manipulation of the joints on the part of the subject. Still, these units in MURPHY's post-learning visual-field population are strongly reminiscent of a relatively small class of cells seen in area 7 that show both visual *and* reach-related responses, dubbed *looking-reaching* neurons by Hyvarinen & Poranen [94]. These neurons responded to a visual stimulus *in*, or a reach *to*, the same region of extrapersonal space, and not to other regions. In contrast to MURPHY's compound visual-joint units however, these units appear to encode location in visual space by direction of gaze rather than by retinal location, possibly reflecting the fact that primates usually fixate what they are reaching for.

Possible Discrepancies, Loose Ends. This latter distinction may in fact represent a fundamental difference between the visual coding strategy used by MURPHY and that used in the brain for predictive arm-motion planning. The lack of precise topographical visual representation amongst the several classes of visually responsive neurons of the posterior parietal cortex could be one indicator of such a difference. It is crucial to note that primates must cope with the transformations of the visual field introduced by head and eye movements, and that they usually foveate the objects to which they reach. In this situation, the retinal image of the arm during reaching is relatively more constant across target locations in comparison to a fixed eye-head system such as MURPHY; what changes instead is the direction of gaze (see [5, 215] for discussion of the issue of head vs. retinal coordinates, and the possible role of the IPL in this transformation). Still, in order to create visual mental images of the arm in space with sufficient resolution as to allow motion planning, it remains theoretically necessary for locality information to be retained in the visual representation. Receptive fields of parietal visual neurons (PVN's) are large (e.g. 10's of degrees in diameter) and often bilateral, but theoretical studies have demonstrated that large receptive fields do not necessarily lead to a loss of locality information [80]. Further, the scale of visual receptive field should reflect the scale of stimulus with which it is normally concerned; the image of the hand and arm during typical reaching trials also subtend in the 10's of degrees.

Also unlike the neurons in MURPHY's primary visual representation, PVN's are highly motion and direction sensitive; however as long as they retain some degree of position information (on the spatial scales of typical stimuli), they can serve to encode the spatial state of the visual field in addition to its instantaneous rate of change. Further questions remain as to how MURPHY's other visual functions might be carried out in the IPL, such as the detection of collisions (visual overlap) between arm image and obstacles during putative mental search. While these ancillary visual functions were kept as algorithmically simple as possible (to simplify later attempts to justify them biologically), proposals as to their actual neural implementation would be pure speculation at this point. One of the most obvious differences between MURPHY's unit populations and, for example, those of the posterior parietal cortex, is that the real cortical areas appear to be rather more heterogeneous in terms of single-unit response properties. Thus, areas 5 and 7 each contain several distinct types of neurons, whose relative concentrations have been shown in some cases to vary gradually across the surface of the cortex. It may be postulated that these cortical areas are actually several distinct *virtual* areas whose units are physically intermixed in the same region of cortex. This type of packing arrangement causes no theoretical problem for the associative synaptic learning algorithm discussed in chapter 3: i) mixing of unit types in the *post*-synaptic population is allowed since the rules governing synaptic formation and response behavior of a given post-synaptic neuron do not depend on the activity of its neighbors (which could be of a different class), and ii) mixing of unit types in the *pre*-synaptic population is allowed since post-synaptic neurons can develop synapses from pre-synaptic units regardless of their class. The disruption of topography in the visual representation would, however, be problematic, since the ability to envision and evaluate the (topographic) spatial relations of arm relative to environmental targets and obstacles is critical to the current theory of motion planning.

Summary. It is eminently clear that no simple one-to-one correspondence holds between MURPHY's monofunctional areas and those of the cerebral cortex involved in the control of visually-guided reaching. The evidence reviewed here does, however, suggest grounds for a functional analogy between MURPHY's forward-kinematic planning subsystem and the "extended internally dependent, predictive or *projectional* action system of the cerebral cortex" [66], including the SMA, and areas 5 and parts of 7 of the posterior parietal association cortex. Primary features of correspondence are:

1. The SMA is active during motor behavior and imagined motor behavior, in possible analogy to MURPHY's sequential controller.

2. The SMA projects to area 5, a body representation that is rich in units encoding joint position, in analogy with MURPHY's

joint-angle population. Area 5, like the SMA, is active during both real and imagined motor behavior, *but only when the action is relative to coordinates of the external environment.*

3. Area 5 projects to area 7, an area rich in visual units and units displaying a variety of combined visual, attentional, and motivational responses. A small number of units have been found to have conjunctive visual and reach-related response properties of exactly the type that emerge post-training in MURPHY's principal visual representation. Area 7 has been implicated in a variety of generic mental imagery functions, and its integrity is known to be critical for the control of visually-guided arm movements.

Within this simplified picture, the SMA is continuously active during motion planning, driving the body representation in area 5 through a sequence of imagined postures. Activity in area 5 in turn drives a corresponding sequence of mental images in area 7 that encode the *visual* consequences of these postures. Throughout the process, these visual mental images are continually monitored and evaluated by unspecified higher brain centers, possibly in the prefrontal association areas to which area 7 projects [97].

A Visual-Feedback Neural Subsystem

To reiterate briefly, Goldberg [66] has elaborated a theory that distinguishes two modes of visually-guided reach control, one associated with the "medial" SMA-based system discussed above,

> "capable of using a model- or hypothesis-driven, feed-forward, predictive mode of control which may rely on efference copy for internal error correction and internal monitoring as the movement unfolds",

and the "lateral" PMA-based system that, in contrast,

> "is dependent on sensory feedback, particularly from foveal vision, and it operates using a corrective, input- or data-driven, closed-loop responsive mode of control."

Thus, the lateral subsystem associated with the PMA seems to dominate in tasks requiring an arm to swoop out reactively to an identified, motivationally significant target, in contrast to control of complex navigational arm movements associated with reaching in the presence of obstacles. Paillard et al. [158] have shown psychophysically that visual guidance of reaching movements is controlled by two additive processes, one based on visual error feedback that operates in the peripheral visual field, the other a positional control system predominant in central vision[2]. According to Mountcastle et al. [153], 60% of the PVN's seen in area 7a exhibit *foveal sparing*, meaning that these neurons can be driven by stimuli in both left and right hemifields, but not in the center of vision. They suggest on this basis that neurons with foveal sparing may underlie the error-feedback mode of control in the visual periphery. Similar reasoning would suggest *non*-foveal-sparing PVN's may underlie the projectional, posture-based style of control discussed in previous sections, which, this evidence would suggest, may be confined to higher-acuity central vision.

We here briefly review the key components of MURPHY's visual-feedback control mode, once again with the intent to lend structure to our interpretations of neural data. The impetus for MURPHY's visual-feedback control system derived from the clear importance of the ability to drive the arm in a straight line to a visual target of interest. Without such an ability, i.e., with only a general-purpose planning mechanism, MURPHY was doomed to mentally flail (i.e., search) his way to targets, even when the targets were entirely unobstructed. Fortunately, a well-defined mathematical relationship exists between the state of the arm, a desired visual-field direction, and the joint commands necessary to push the hand in the desired direction[3]. Moreover, these visual and limb-state variables were naturally encodable in a neurally-plausible style of representation. In addition to

[2]This introduces an apparent contradiction, since the study of Paillard et al. implicates *peripheral* vision in visual-feedback control, whereas Goldberg [66], as above, suggests a primary role for *foveal* vision in visual-feedback control. Since the neural data regarding the detailed nature of visual responses in the PMA is virtually nonexistent, resolution of this conundrum awaits further data.

[3]This relationship is the inverse differential kinematics of the arm, traditionally expressed as an inverse Jacobian matrix; see [39, 159] for background.

his visual-field and joint-angle populations therefore, MURPHY was endowed with i) a population of visual direction-selective units that were sensitive to motion of the hand through the visual field independent of its retinal location (i.e., with full-visual-field receptive fields), and ii) a population of joint-velocity units that explicitly encoded direction and magnitude of change in the positions of the joints. The functional mapping used for visual feedback control of the arm was then implemented with two input populations, namely the joint-angle and visual-direction populations, both simultaneously projecting to the joint-velocity population (fig. 2.2). Activity induced in this latter population specifies the appropriate degree of flexion or extension to each of the joints as needed to nudge the hand in the direction of a given visual-field target. In analogy with control modes in the oculomotor system, these commanded joint perturbations can be thought of as *joint saccades* or, as the movements become infinitesimal, error-driven *smooth pursuit* movements. It was pointed out in section 4 that the usefulness of such a control system by itself is greatly diminished when obstacles are present in the workspace, since a tendency to plunge the arm straight towards a visual target can in this case be a liability.

A fascinating bit of evidence that accentuates the distinction between internally planned movement and feedback-based control is reported by Moll & Kuypers [147], who demonstrated that monkeys with premotor lesions will reach directly and accurately towards food even when obstructed by a transparent barrier, i.e., will fail to plan the appropriate circuitous trajectory. Unfortunately, the lesions reported in this study included both the SMA and PMA, so that their individual contributions to this deficit are not known. Nonetheless, the evidence is strongly supportive of the notion that normal limb control in primates is achieved through the transparent superposition of at least two modes of control—one reactive, one planned. From this perspective, the pathology observed in these experiments, which has usually been described as "a release of grasping reflexes", may be more appropriately viewed as the abolishment of the on-line predictive planning faculty that normally averts impending collisions of the hand and arm. It would be similarly illuminating if due to brain injury or experimental lesion a subject were observed to *retain* the ability to execute complex pre-planned motions of the arm relative to objects in surrounding space, while having *lost* the ability to execute rapid visually-directed reaching.

Role of the PMA and Associated Areas. A principal basis for the distinction between medial (SMA-based) and lateral (PMA-based) premotor subsystems has been their pattern of anatomical connections. Of current concern is the special relationship between the PMA and areas of the parietal cortex: whereas the SMA has major connections to area 5 whose possible function is discussed above, the PMA receives somatosensory projections from S2 (the major cortical site that provides limb-state information to area 5), and combined visual and somatic inputs from the IPL (see [66] for greater anatomical detail). Since we seek evidence for a premotor area capable of issuing joint saccades that is conjointly enervated by i) a static body-image population encoding current limb-posture, and ii) a direction-selective visual population with very large receptive fields, the PMA seems an ideal candidate. Unfortunately, the PMA is the most poorly understood premotor area, and very little direct evidence is yet available as to its specific motor-related functions. An issue of theoretical importance to the current discussion regards the type of motor code explicit in PMA neurons: are they more akin to the postural motor correlates of SMA neurons, or to the phasic single-muscle "saccades" induced by stimulation of cells in M1? Electrocortical stimulation studies are ambivalent on this question (see [92]). A study of the responses of single neurons in the PMA uncovered at least one class of neurons that fired differentially in relation to wrist flexion and extension in a visual-motor tracking task [114], but the precise relationship between cell activity and motor parameters was not determined. In this same study, Kubota & Hamada [114] also found neurons responsive to visual stimuli with shorter latencies than movement-related neurons, implying that visual information is at least available in the PMA in advance of movement, and could therefore play a role in the control of movement. A potential problem is that the direct visually-guided reaching behavior under consideration here is not abolished with complete ablation of premotor cortex [147]. It is possible, however, that the projections from the IPL directly to subcortical motor systems, such as the deep cerebellar nuclei, could to some degree act in parallel to those interconnecting the IPL with the PMA transcortically (see [66]). The existence of multiple neural subsystems performing the same (or highly similar) function is by no means unprecedented, a notable example being the control of saccadic eye movements by both the superior colliculus and the frontal eye fields. Why not multiple parallel control of visually-guided joint saccades as well?

Summary. Once again it is clear that the analogy between MURPHY's visual-feedback mode of control and the lateral, PMA-based premotor system of the primate brain is inconclusive, but suggestive. It seems clear that the PMA cannot, in any case, be the *only* channel for visually-driven joint saccades, since its ablation leaves feedback-driven visual reaching intact. However, the following would seem to implicate the PMA as at least a partial contributor to this mode of limb control: i) both limb-state and visual-direction information are known with relative certainty to be represented in the parietal cortex in areas 2 and 7, respectively, in a form strongly resembling the style of connectionist representation used by MURPHY's joint-state and direction-selective visual populations, and ii) both of these populations project to the PMA, an area implicated through evolutionary, anatomical, physiological, and neuropathological arguments (see [92, 66, 214]) as a central premotor component in a feedback-mediated (as opposed to *predictive*) subsystem for the control of visually-guided reaching. With MURPHY's counterpart subsystem as a reference point, it does not seem unreasonable to infer that this pathway could learn and ultimately generate the pattern of joint saccades (or the smooth, infinitesimal equivalents) needed to push the hand towards a visual target at the center of gaze. It remains to be established experimentally whether or not this kind of role for the PMA is plausible.

Summary of Biological Issues

Much of MURPHY's usefulness as a neurobiological tool has come in providing concrete working principles that demonstrate how known areas of the cerebral cortex could interact to carry out an interesting and difficult visual-motor control function. Thus, the emphasis in this work has been on the exploration of possible *systems-level* functions of the cerebral cortex. We once again summarize with a list of questions upon which the foregoing analysis has shed light:

1. How a number of independent bits of evidence strongly suggest that the brain's internal language for limb motion planning is primarily concerned with posture and kinematics rather than with forces and limb/load dynamics (this is *not* to say that the brain cannot or does not control limb actions that involve accurate load management and force and velocity control, but that these aspects of movement control are not readily simulated within the cortical "mental model"). Cited evidence included i) the *mass-spring* nature of the muscle interface, which for lightly loaded movements is essentially a form of robust position control, ii) the high proportion of position-related joint neurons in area 5, an area intimately involved in the planning and control of limb movements in extrapersonal space, iii) the effects of electrocortical stimulation to both the SMA and area 5 which tend to produce slowly habituating limb postures, rather than the phasic, twitch-like responses of stimulation to primary motor cortex, iv) the indications that *mental practice* is most effective for kinematics-intensive motor skills, rather than those that involve high speeds and varying loads, and v) the fact that amputees subjectively perceive the *position* but not motion of their phantom limbs in response to issued motor commands, suggesting that the corollary discharge associated with intended movement encodes postural rather than motion-related limb information.

2. How and why the cortical representations of limb posture vs. limb movement may play distinct roles in programmed-feedforward vs. visual feedback modes of limb control. Moreover, how very large direction-selective visual receptive fields, as seen in the posterior parietal cortex (and in analogy to units in MURPHY's hand-direction population), are ideally suited to drive *joint saccades* in a visual-feedback limb controller.

3. How a neural premotor area such as the SMA, known simply to be involved in "motor programming", may specifically act via corollary discharge to drive a posterior parietal mental model that relates limb postures to their predicted visual-field consequences. Also, how such a faculty could develop from scratch through sensory-motor learning.

4. How the concept of *receptive field* may be confusing in describing the functions of single neurons involved in integrated sensory-motor function. As is the case for the complex visual-motor neurons that develop in MURPHY's visual field population during learning (and that closely resemble the class of *arm projection* neurons found in area 7), the presence of joint-related input to an otherwise visual neuron, rather than indicating a simple joint-sensory or motor representational role, may signal the locus of a learned visual-kinematic mapping, i.e., the site at which a joint-command corollary discharge gives rise to a predictive visual image.

5. How, where and why visual mental imagery might be specifically used in the brain for the programming and control of limb (and other) body movements. As a corollary, how the brain's ubiquitous topographically-mapped visual representations can lead to cheap solutions to difficult, highly visual problems.

6. Finally, as we have seen in chapter 3, how the *sigma-pi* unit may be a biologically-significant model neuron. Acting as an interpolating multidimensional lookup-table, the *sigma-pi* unit can effect fast, single-layer synaptic learning of ecologically-relevant smooth nonlinear functions, using only simple Hebb-type biophysical mechanisms.

Chapter 8

Conclusions

Scientific and Engineering Lessons Learned

Reaching for visual targets in the presence of obstacles is a classically hard problem. This principal goal of this work has been to demonstrate a new architectural approach to the problem that takes advantage of the possibility of learning from active experience, the strengths of neurally-inspired connectionist representations, and the utility of heuristic search when *good* performance, even if suboptimal, is adequate for the task. Along these lines, the four principal messages of this book are discussed below.

Learning by Doing

This under-exploited learning paradigm is a powerful means by which a learning system can extract information from an unknown reactive environment, particularly when an accurate advance model of system/environment interaction is not available, or when either the system or the environment is expected to change over time. Given that this active style of learning is known to be of foundational importance in the cognitive development of human infants, it seems likely to be a technique of great utility in future AI and machine learning systems.

Learning by doing is useful in the wide range of real-world environments where i) a systematic relationship exists between the learning agent's outgoing commands and the incoming stream of sensory information, and ii) the ability to internally manipulate these relationships is of ecological utility to the learner. Also, the learning agent can in many cases learn and use these input-output relationships in both directions simultaneously. For example, MURPHY learns and uses the "forward" mapping between limb-posture commands and incoming visual-field information to run internal visual-motor simulations. He also learns and uses the "inverse" mapping between a desired visual-field direction and the corresponding joint-saccade commands, useful in driving his hand straight to a visual target.

Sensory-motor domains are thus the most obvious candidates for learning by doing, since the relevant laws and interrelationships are highly reliable, and of great utility to intelligent organisms. Beyond the control of reaching, the intelligent visual control of locomotion is another ripe domain of application for these techniques. The problem of speech production and comprehension has also been suggested as ideal ground for learning by doing [180]. Perhaps the most interesting possibilities, however, arise when the concepts of "motor command" and "sensory input" are relaxed to allow for learning in an environment that is entirely *internal* to the learning agent. Thus, "command" states, rather than being confined to the simple control of motor acts, may instead be allowed to control processes (i.e., the "motors" of the mind!). For example, attentional processes are of this general character, where outgoing, or command, activity is allowed to modulate the functioning of purely internal sensory systems. Likewise, "sensation" may be relaxed to include internal state that is not directly tied to the peripheral sensory epithelia, such as the internal, affective state of the learning agent. Thus, one could imagine an agent that learns the relationship between commanded changes in either pattern or level of attention, and the resulting changes in the agent's affective state (if any). Scenarios of this kind abound, limited only by the imagination of the system architect.

Learning with Lookup-Tables

Much emphasis has been given in this work on the virtues of locally-smoothing multi-dimensional lookup tables for nonlinear function learning, which may be represented either neurally (fig. 3.4, also

[207, 143]), or with other more traditional data structures [4, 149, 155, 166]. The standard advantages of tabular learning are speed, algorithmic simplicity, and representational clarity. In addition, tabular learning is particularly appropriate for cases where little or no advance knowledge as to the structure of the mappings to be learned is available (e.g., linearity). The foremost disadvantage of tabular learning derives from the fact that generalization is by definition local, such that what is learned from any single training example generalizes only through a local neighborhood in the input space. Thus, it is necessary to "see" training examples drawn from across all relevant (i.e., useful and/or legal) regions in the domain space. In general, the smoother the function to be learned, the larger may be the neighborhood of valid generalization surrounding a single training input.

Tabular learning is thus limited to functional mappings in which the total amount of data to be learned is tractable. In the worst case the amount of needed data increases exponentially with the dimension of the input space. In many cases, however, the volume of the multidimensional input space that is actually used by the learning agent may remain small. For example, while a robot arm can have 7 degrees of freedom, it may be the case that only small bounded subregions of the 7-d joint space are ever used in practice, leaving vast regions of the joint space entirely unvisited. The amount of input data needed to approximate the 7-d mapping is therefore correspondingly reduced, since it need only represent the relatively small bounded subregions of joint space that are of behavioral significance. Moreover, for *smooth* functional mappings, many fewer training examples are needed to maintain a given level of accuracy in locally-generalizing learning schemes [155]. For discontinuous learning problems of very high dimension, such as the association of names with visual images of objects, a tabular learning scheme simply memorizes examples until table capacity is exceeded (chapter 3).

In sum, the implementation costs of tabular learning can in principle grow exponentially with the dimensionality of the input space. On the other hand, many real-world functions that are of relevance to intelligent organisms may be sufficiently smooth and of bounded domain that this "curse of dimensionality" is significantly mitigated.

The Connectionist Architecture

MURPHY's style of architecture is ideal for an important class of biologically-relevant learning problems. At a global systems level, this type of architecture allows a learning agent to simultaneously acquire any number of functional mappings among the internally represented sensory, motor, and affective state spaces—all without the need for intelligent external supervision. The architecture is biologically advantageous since its power can be increased by simply proliferating unit populations and their interconnecting pathways, or by elaborating the representational content of existing populations. Interestingly, these measures are central among those used to chart the evolution of the mammalian cerebral cortex.

At the heart of a learning agent, a connectionist architecture is a natural and uniform way of allowing a variety of internally represented sensory, motor, or other more abstract quantities to interact, since learning reduces to synaptic weight modification, and performance reduces to the propagation of activation levels among simple units [53, 181]. In the context of sensory-motor learning problems such as the visual control of limb movement, the use of highly decoded connectionist representations can lead to another important advantage. For example, MURPHY's highly decoded full-visual-field representation is ideally suited for the rapid parallel computation of visually-intensive functions, such as collision detection, that are critical for the planning and control of limb movements. A entire class of comparable problems may be imagined in which the image-like character of a connectionist representation could lead to efficient implementation of algorithms for which space and topography are critical. Not limited to vision, the generation and manipulation of topographic auditory and body-state imagery is equally possible.

As the promise of connectionism has been in part a technological one, it bears mention as well that an architecture like MURPHY's is also a sensible way to build a very fast computer from unreliable components, since it is both massively parallel and fault tolerant. With continuing advances in VLSI and optical technology (e.g., [90, 133]), it should be feasible in the not-too-distant future to build compact real-time vision-based robotic control systems.

Mental Models, Heuristic Search

Where optimality of performance is not crucial, simple heuristic search can be successfully used in conjunction with a mental model to solve classically difficult computational spatial reasoning problems, such as inverse kinematics and motion planning for a multi-link robot arm. Where MURPHY uses on the order of *hundreds* of parallel virtual-machine operations in planning complex paths for his 3-*dof* arm, traditional serial approaches that emphasize path optimality consume in the tens or hundreds of *millions* of serial machine operations. Moreover, MURPHY proceeds without equations, complex multi-phase algorithms, or the general purpose computers upon which they depend.

To summarize, the general approach outlined in this work has entailed i) the adoption, where possible, of a parallel rather than serial machine model in order to significantly speed up traditionally expensive operations, and ii) the relaxation of optimality as a primary performance criterion, in favor of heuristic goodness and algorithmic simplicity. One wonders whether other classically difficult AI-relevant problems, such as the control of deductive and inductive logical inference, could be recast in a similar way.

Reflections on Brain and Behavior

A more abstract goal of this work has been to demonstrate that, when approaching an ecologically-relevant computational problem, armed to the hilt with neurally- and psychologically-inspired elemental mechanisms, common-sense design principles can lead to engineering solutions that are related to those underlying the counterpart biological information processes. In concluding sections of chapters 6 and 7 we noted a number of ways in which MURPHY's design has "co-evolved" with the brain, presumably in response to similar computational pressures. The most salient example of this process was the division of labor between programmed-feedforward and visual-feedback modes of limb control.

Perhaps more important has been MURPHY's role as a conceptual tool in interpreting aspects of the psychological and neural data.

The act of having built a working system under many of the same computational constraints as are faced by the brain, *even if wrong in the specifics of neural implementation or psychological process*, nonetheless enables an extra degree of discrimination in the attempt to interpret a large and complex body of scientific literature. One example is in the interpretation of single-neuron response properties: *Gedanken* electrophysiology on MURPHY's visual-field population provides a conceptual template against which to compare the conjoint visual/limb-related unit responses seen in the posterior parietal association cortex. Compelling similarities were seen in this comparison, justifying a transfer of functional interpretation from software to wetware.

A final way in which an engineered system, such as MURPHY, can be of use as a biological or psychological tool is by concrete *example*. Thus, the concepts of *visual mental imagery*, *corollary discharge*, and *motor programming*, among others, have been much discussed by behavioral scientists, but concrete proposals as to their neural implementations have been conspicuously few in number. Concrete examples of *possible* solutions often make for more directed and successful searches for the truth.

Future Directions

Pragmatic Extensions

A number of pragmatic extensions to MURPHY could make the approach of true industrial significance. These include i) extension to a 6-*dof* arm with 3-d vision system as discussed in chapter 4, ii) more intelligent search heuristics (e.g., ruling out the obviously impossible paths that currently occupy much of MURPHY's mental efforts, iii) more intelligent quantization of search—i.e., searching in broad strokes in unencumbered parts of the workspace and finer strokes in "tighter squeezes", iv) two parallel visual systems that would allow the separation of finer pattern-vision (used to detect targets and obstacles) from coarser arm-vision (i.e., that which is used to represent the much larger arm, both visually and mentally)—this division of labor would eliminate an unnecessary design tension that currently exists, and v) development of a truly parallel implementation that would support absolutely real-time control of a 6-*dof* arm in a complicated workspace. Figure 8.1 shows the user interface to a simulation that has been written to test the feasibility of extending MURPHY's algorithms to more complex arms (up to 7-*dofs*) under the guidance of a binocular vision system.

Two other areas of relatively immediate exploration would be i) *compliant* control, in which MURPHY could use his visual-feedback control mode to apply directed forces to surfaces in his workspace, and ii) independent partioning of motion planning for gross movements of the arm vs. fine movements of the hand, involving separately learned visual-kinematic maps for each.

Theoretical Extensions

A number of visual-motor learning problems differ in an important respect from the control of limb movements relative to the body, in that they involve the prediction of visual states of the *environment*, in addition to, or instead of, prediction of visual states of only the limb(s). For example, MURPHY is only able to predict the visual appearance of his own arm as it moves through the workspace, but would be unable, within his current architecture, to predict the visual trajectory of an autonomously moving object in his visual field, such as another arm, or even that of an object he himself were carrying. For problems of visually-guided *locomotion* through a complex environment, it is crucial to predict the self-induced flow of features through the visual field, in order to differentiate self-induced flow from true external motion in the environment. Prediction of this kind requires a more sophisticated mental model than has MURPHY, since predictions must take into account not only the current state of the body, but also the current state of objects in the visual environment. An initial attack on a problem of this kind was reported in previous work [134], which described a connectionist system that learned to mentally rotate simulated 3-d wire-frame objects.

A second major issue that remains to be addressed, and that was also raised more directly in [134], is that of the connectionist implementation of sequential control. Currently, MURPHY's sequential

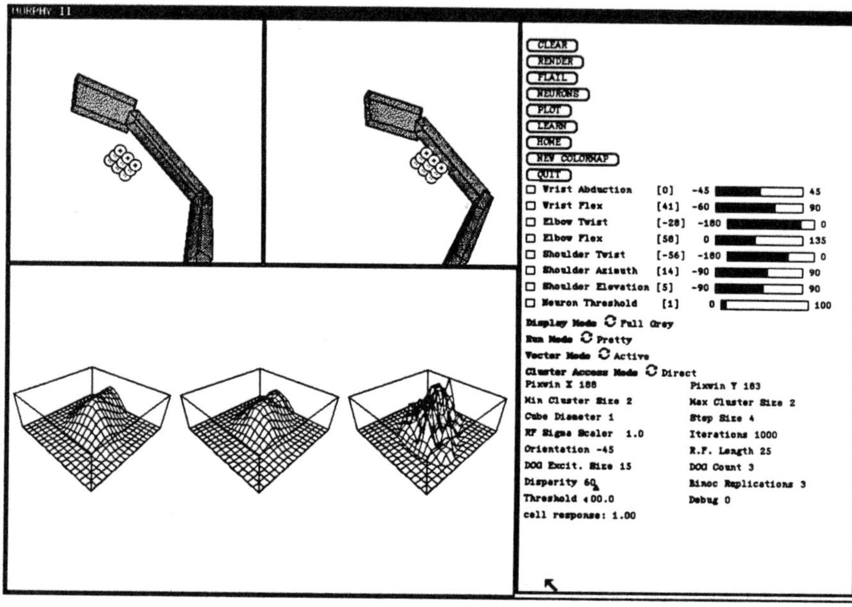

Figure 8.1: Simulation The upper frames at left show the arm from the perspective of right and left eyes. An oriented binocular receptive field is also shown. Leftmost plot shows the unconditioned output of a visual unit (receptive field not shown) over two joint degrees of freedom. Center plot shows the approximation to this function as learned from examples. Right plot shows the mean-squared error (x 10).

controller remains a black box with no commitment to a particular underlying architecture, connectionist or otherwise. The ability to systematically sequence through mental states is clearly central to intelligent (and therefore brain) function, and its architectural underpinnings deserve to be explained.

A final, still more difficult, unresolved issue is that of higher-level representation and learning. Nowhere in MURPHY's architecture is there room for knowledge or learning *about* reaching for targets, such as, "Gee whiz, I remember I had trouble getting my wrist under that long skinny obstacle last time because my elbow kept bumping into it—maybe I should try going over the *top* this time...!". Knowledge of this type is crucial for truly intelligent performance in general, and for visually-guided reaching in particular. It would therefore be of great interest to explore neurally-plausible representational substrates for this more symbolic type of reasoning process as well.

Bibliography

[1] Ackley, D.H. Associative learning via inhibitory search. Presented at the 1988 Snowbird Conf. on Neural Networks for Computing.

[2] Ackley, D.H., Hinton, G.E., & Sejnowski, T.S. A learning algorithm for Boltzmann machines. *Cognitive Science* , 1985, *9*, 147-169.

[3] Albus, J.S. A theory of cerebellar function. *Math. Biosci.*, 1971, *10*, 25-61.

[4] Albus, J.S. A new approach to manipulator control: the cerebellar model articulation controller (CMAC). *ASME J. of Dynamic Systems, Measurement, & Control*, September 1975, 220-227.

[5] Andersen, R.A. Inferior parietal lobule function in spatial perception and visuomotor integration. In *Handbook of Physiology, Sec. 1, The Nervous System (vol. 5)*, F. Plum, (Ed.), Bethesda, MD: American Physiological Society, 1987.

[6] Andrews, J.R. & Hogan, N. Impedance control as a framework for implementing obstacle avoidance in a manipulator. In *Control of manufacturing processes and robotic systems.* ASME: New York, 1983.

[7] Artola, A. & Singer, W. Long-term potentiation and NMDA receptors in rat visual cortex. *Nature*, 1987, *330*, 649-652.

[8] Asatryan, D.G. & Fel'dman, A.G. Biophysics of complex systems and mathematical models. Functional tuning of nervous system with control of movement or maintenance of a steady posture: I. Mechanographic analysis of the work of the joint on execution of a postural task. *Biophysics*, 1965, *10*, 925-935.

[9] Atkeson, C.G. Roles of knowledge in motor learning. Technical Report 942, MIT AI Lab, 1987.

[10] Atkeson, C.G. & Reinkensmeyer, D.J. Using associative content-addressible memories to control robots. A.I. Memo 1124, C.B.I.P. Paper 37, Massachussets Institute of Technology, 1989.

[11] Atkeson, C.G. & Hollerbach, J.M. Kinematic features of unrestrained vertical arm movements. *J. Neurosci.*, 1985, *5*, 2318-2330.

[12] Baldi, P. & Heiligengerg, W. How sensory maps could enhance resolution through ordered arrangements of broadly tuned receptors. *Biol. Cybern.*, 1988, *59*, 313-318.

[13] Ballard, D.H. Cortical connections and parallel processing: structure and function. *Behav. Brain Sci.*, 1986, *9*, 67-120.

[14] Ballard, D.H. Interpolation coding: a representation for numbers in neural models. *Biol. Cybern.*, 1987, *57*, 389-402.

[15] Ballard, D.H., Hinton, G.E., & Sejnowski, T.J. Parallel visual computation. *Nature*, 1983, *306*, 21-26.

[16] Barraquand, J. Langlois, B., & Latombe, J. Numerical potential field techniques for robot path planning. Report No. STAN-CS-89-1285, Department of Computer Science, Stanford University, 1989.

[17] Barron, A.R. Predicted squared error: a criterion for automatic model selection. In *Self-organizing methods in modeling: GMDH-type algorithms*, S.J. Farlow, (Ed.), New York: Marcel Dekker, 1984.

[18] Barron, A.R. & Barron, R.L. Statistical learning networks: a unifying view. Presented at the *1988 Symposium on the Interface: Statistics and Computing Science*, Reston, Virginia.

[19] Barto, A.G. & Sutton, R.S. Goal-seeking components for adaptive intelligence: an initial assessment. AFWAL-TR-81-1070, Avionics Laboratory, Air Force Wright Aeronautical Laboratories, Wright-Patterson AFB, Ohio 45433, 1981.

[20] Barto, A.G., Sutton, R.S., & Anderson, C.W. Neuron-like adaptive elements that can solve difficult learning control problems. *IEEE Trans. on Sys., Man, Cybern.*, 1983, *SMC-13*, 834-846.

[21] Bauer, J. & Held, R. Comparison of visually guided reaching in normal and deprived infant monkeys. *J. Exp. Psychol.: Animal Behavior Processes*, 1975, *1*, 298-308.

[22] Becker, J.D. A model for the encoding of experiential information. In *Computer models for thought and language.*, R. Schank & K.M. Colby, (Eds.), San Francisco: Freeman, 1973.

[23] Bell, C.C. Duration of plastic change in a modifiable efference copy. *Brain Res.*, 1986, *369*, 29-36.

[24] Bentley, J.L. Multidimensional binary search trees used for associative searching. *CACM*, 1975, *18*, 509-517.

[25] Bernstein, N.A. *The co-ordination and regulation of movements.* Pergamon Press, 1967.

[26] Bizzi, E., Accornero, N., Chapple, W., & Hogan, N. Posture control and trajectory formation during arm movement. *J. Neurosci.*, 1984, *4*, 2738-2744.

[27] Bliss, T.V.P. & Lømo, T. Long-lasting potentiation of synaptic transmission in the dentate area of the anaesthetized rabbit following stimulation of the perforant path. *J. Physiol.*, 1973, *232*, 331-356.

[28] Brodmann, K. [Principles of comparative localization in the cerebral cortex presented on the basis of cytoarchitecture.] Leipzig: Barth, 1909.

[29] Brooks, R.A. Solving the find-path problem by good representation of free space. *IEEE Trans. on Sys., Man, Cybern.*, 1983, *SMC-13*, 190-197.

[30] Brooks, R.A. Planning collision-free motions for pick-and-place operations, *Int. J. Robotics Res.*, 1983, *2*, 19-40.

[31] Broomhead, D.S. & Lowe, D. Multivariable functional interpolation and adaptive networks. *Complex Systems*, 1988, *2*, 321-355.

[32] Brown, T.H., Chapman, P.F., Kairiss, E.W., & Keenan, C.L. Long-term synaptic potentiation. *Science*, 1988, *242*, 724-728.

[33] Brown, T.H., Chang, V.C., Ganong, A.H., Keenan, C.L., & Kelso, S.R. Biophysical properties of dendrites and spines that may control the induction and expression of long-term synaptic potentiaion. In *Long-term potentiation,: from biophysics to behavior.* Alan R. Liss, Inc., 1988.

[34] Bruner, J.S. The growth and structure of skill. In *Mechanisms of motor skill development*, K. Connolly, (Ed.), New York: Academic Press, 1970.

[35] Brumbach, W.B. & Beckow, P.A. Effectiveness of mental practice upon the learning of two types of gross motor skills. *Abstracts of Research Papers.* Washington, D.C.: AAHPER, 1968.

[36] Bullock, D. & Grossberg, S. Neural dynamics of planned arm movements: emergent invariants and speed-accuracy properties during trajectory formation. *Psych. Rev.*, 1988, *95*, 49-90.

[37] Carbonell, J.G. & Gil, Y. Learning by experimentation. In *Proc. of the 4th Int. Workshop on Machine Learning*, 1987, 256-266.

[38] Clark, L.V. Effect of mental practice on the development of a certain motor skill. *Research Quarterly*, 1960, *31*, 560-569.

[39] Craig, J.J. *Robotics.* Reading, MA: Addison-Wesley, 1986.

[40] Craik, K. *The nature of explanation.* Cambridge University Press, 1943.

[41] Crick, F.H.C. & Asanuma, C. Certain aspects of the anatomy and physiology of the cerebral cortex. In *Parallel distributed processing: explorations in the microstructure of cognition, vol. 2*, D.E. Rumelhart, J.L. McClelland, (Eds.), Cambridge, MA: Bradford, 1986.

[42] Denker, J.S., Gardner, W.R., Graf, H.P., Henderson, D., Howard, R.E., Hubbard, W., Jackel, L.D., Baird, H.S., & Guyon, I. Neural network recognizers for handwritten zip-code digits. In *Advances in Neural Information Processing Systems vol. 1*, D.S. Touretzsky, (Ed.), San Mateo, CA: Morgan Kaufmann, 1989.

[43] Deuel, R. Loss of motor habits after cortical lesions. *Neuropsychologia*, 1977, *15*, 205-215.

[44] Deutsch, G., Bourbon, W.T., Papanicolaou, A.C., Eisenberg, H.M. Visuospatial tasks compared via activation of regional cerebral blood flow. *Neuropsychologia*, 1988, *26*, 445-452.

[45] Drowatzky, J.N. *Motor learning: principles and practices.* Minneapolis: Burgess, 1975.

[46] Durbin, R. & Rumelhart, D.E. Product units: a computationally powerful and biologically plausible extension to backpropagation networks. *Neural Computation*, 1989, *1*, 133.

[47] Elsley, R.K. A learning architecture for control based on back-propagation neural networks. 1988 Snowbird Meeting on Neural Networks for Computing.

[48] Erikson, R.P. On the neural bases of behavior. *American Scientist*, May-June 1984, 233-240.

[49] Farrah, M.J. The neurological basis of mental imagery: a componential analysis. In *Visual cognition*, S. Pinker, (Ed.), Cambridge, MA: MIT Press, 1984.

[50] Farrah, M.J. Is visual imagery really visual? Overlooked evidence from neuropsychology. *Psych. Rev.*, 1988, *95*, 307-317.

[51] Faverjon, B. Obstacle avoidance using an octree in the configuration space of a manipulator. In *Proc. IEEE Int. Conf. Robotics and Automation*, Atlanta, March 1984.

[52] Faverjon, B. Hierarchical object models for efficient anti-collision algorithms. In *Proc. IEEE Int. Conf. Robotics and Automation*, 1988, 333-340.

[53] Feldman, J.A. & Ballard, D.H. Connectionist models and their properties. *Cognitive Science*, 1982, *6*, 205-254.

[54] Fel'dman, A.G. Once more on the equilibrium point hypothesis (model) for motor control. *J. Motor Behavior*, 1986, *18*, 17-54.

[55] Finke, R.A. Theories relating mental imagery to perception. *Psych. Bulletin*, 1985, *98*, 236-259.

[56] Finke, R.A., Kosslyn, S.M. Mental imagery acuity in the peripheral visual field. *J. Exp. Psychol., Human Perception and Performance*, 1980, *6*, 244-264.

[57] Finke, R.A. & Shepard, R.N. Visual functions of mental imagery. In *Handbook of perception and human performance*, L. Kaufman & J. Thomas, (Eds.), New York: Wiley, 1985.

[58] Finkel, L.H. & Edelman, G.M. Population rules for synapses in networks. In *Synaptic function*, G.E. Edelman, W.F. Gall, & W.M. Cowan, (Eds.), Wiley, 1987.

[59] Gamble, E. & Koch, C. The dynamics of free calcium in dendritic spines in response to repetitive synaptic input. *Science*, 1987, *236*, 1311-1315.

[60] Gardner, E.P. Somatosensory cortical mechanims of feature detection in tactile and kinesthetic discrimination. *Can. J. Physiol. Pharmacol.*, 1988, *66*, 439-454.

[61] Georgopoulos, A.P., Kalaska, J.F., Caminiti, R., & Massey, J.T. On the relations between the direction of two-dimensional arm movements and cell discharge in primate motor cortex. *J. Neurosci.*, 1982, *2*, 1527-1537.

[62] Georgopoulos, A.P., Lurito, J.T., Petrides, M., Schwartz, A.B., & Massey, J.T. Mental rotation of the neuronal population vector. *Science*, 1989, *243*, 234-236.

[63] Giles, C.L. & Maxwell, T. Learning, invariance, and generalization in high-order neural networks. *Applied Optics*, 1987, *26*, 4972-4978.

[64] Gilmore, R.L, & Storlurow, L.M. Motor and "mental" practice of a ball and socket task. *American Psychology*, 1951, *6*, 295.

[65] Ginsburg, H. & Opper, S. *Piaget's theory of intellectual development.*, New Jersey: Prentice Hall, 1969.

[66] Goldberg, G. Supplementary motor area structure and function: review and hypotheses. *Behav. Brain. Sci.*, 1985, *8*, 567-616.

[67] Grossberg, S. Adaptive pattern classification and universal recoding. *Biol. Cybern.*, 1976, *23*, 121-134.

[68] Guez, C. Voluntary movement. In *Principles of neural science*, E.R. Kandel & J.H. Schwartz, (Eds.), New York: Elsevier, 1985.

[69] Hebb, D.O. *The organization of behavior.* New York: Wiley, 1949.

[70] Hein, A. & Held, R. Dissociation of the visual placing response into elicited and guided components. *Science*, 1967, *158*, 390-392.

[71] Held, R. & Hein, A. Movement-produced stimulation in the development of visually-guided behavior. *J. Comp. and Physiol. Psych.*, 1963, *56*, 872-876.

[72] Held, R. Plasticity in sensory-motor systems. In *Readings from Scientific American, Perception: Mechanisms and Models*, San Francisco: W.H. Freeman. (article originally published in *Scientific American*, Nov. 1965).

[73] Held, R. & Bauer, J.A. Development of sensorially-guided reaching in infant monkeys. *Brain Res.*, 1974, *71*, 265-271.

[74] Henry, F.M. & Rogers, D.E. Increased response latency for complicated movements and a "memory drum" theory of neuromotor control. *Research Quarterly*, 1960, *31*, 448-458.

[75] Hildreth, E.C. & Hollerbach, J.M. Artificial intelligence: computational approach to vision and motor control. In *Handbook of Physiology, Sec. 1 The Nervous System (vol. 5)*, F. Plum, (Ed.), Bethesda, MD: American Physiological Society, 1987.

[76] Hinton, G.E. Some demonstrations of the effects of structural descriptions in mental imagery. *Cognitive Science*, 1979, *3*, 231-250.

[77] Hinton, G.E. Connectionist learning procedures. Technical Report CMU-CS-87-115, Pittsburgh, PA: Carnegie Mellon University, 1987.

[78] Hinton, G.E. & Parsons, L.M. Frames of reference and mental imagery. *Attention and performance IX*, A. Baddeley & J. Long, (Eds.), Hillsdale, NJ: Erlbaum, 1981.

[79] Hinton, G.E. Some computational solutions to Bernstein's problems. In *Human motor actions—Bernstein reassessed.*, H.T.A. Whiting (Ed.), Elseview Science Publishers B.V. (North Holland), 1984.

[80] Hinton, G.E. Distributed representations. In *Parallel distributed processing: explorations in the microstructure of cognition, vol. 1*, D.E. Rumelhart, J.L. McClelland, (Eds.), Cambridge, MA: Bradford, 1986.

[81] Hinton, G.E. & Sejnowski, T.J. Learning and relearning in Boltzman machines. In *Parallel distributed processing: explorations in the microstructure of cognition, vol. 1*, D.E. Rumelhart, J.L. McClelland, (Eds.), Cambridge, MA: Bradford, 1986.

[82] Hofsten, von C. Eye-hand coordination in the newborn. *Developmental Psych.*, 1982, *18*, 450-461.

[83] Hogan, N. An organizing principle for a class of voluntary movements. *J. Neurosci.*, 1984, *4*, 2745-2754.

[84] Holst, E. von & Mittelstaedt, H. The principle of reafference: interactions between the central nervous system and the peripheral organs. In *Perceptual processing: stimulus equivalence and pattern recognition*, P.C. Dodwell (Ed. and trans.), New York: Appleton, 1971. (Reprinted from *Die Naturwissenschaften*, 1950.)

[85] Holland, J.H., Holyoak, K.J., Nisbett, R.E., & Thagard, P.R. *Induction: processes of inference, learning, and discovery.* Cambridge, MA: MIT Press, 1983.

[86] Houk, J.D., Crago, P.E., & Rymer, W.Z. Function of the spindle dynamic response in stiffness regulation—a predictive mechanism provided by nonlinear feedback. In *Muscle receptors and movement*, H. Taylor & A. Prochazka, (Eds.), London: MacMillan, 1981.

[87] Houk, J.C. & Rymer, W.Z. Neural control of muscle length and tension. In *Handbook of Physiology: Sec. 1., Vol. 2, Motor Control*, V.B. Brooks, (Ed.), Bethesda, MD: American Physiological Society, 1981.

[88] Howard, I.P. *Human visual orientation.* New York: Wiley, 1982.

[89] Howden, W.E. The sofa problem. *Comput. J.*, 1968, *11*, 299-301.

[90] Hsu, K., Brady, D., & Psaltis, D. Experimental demonstrations of optical neural computers. *Neural information processing systems*, D.Z. Anderson, (Ed.), New York: American Institute of Physics, 1988.

[91] Hubel, D.H. & Wiesel, T.N. Receptive fields, binocular interaction, and functional architecture in the cat's visual cortex. *J. Physiol.*, 1962, *160*, 106-154.

[92] Humphrey, D.R. On the cortical control of visually directed reaching: contributions by nonprecentral motor areas. In *Posture and movement*, R.E. Talbott & D.R. Humphrey, (Eds.), New York: Raven Press, 1979.

[93] Hwang, Y.K. & Ahuja, N. Path planning using a potential field representation. Coordinated Science Laboratory Technical Report UILU-ENG-88-2251, University of Illinois, 1988.

[94] Hyvärinen, J. & Poranen, A. Function of the parietal associative area 7 revealed from cellular discharges in alert monkey. *Brain*, 1974, *97*, 673-691.

[95] Ito, M. Long-term depression. *Ann. Rev. Neurosci.*, 1989, *12*, 85-102.

[96] James, W. *The principles of psychology.* Cambridge, MA: Harvard University Press, 1983. (originally published in 1890).

[97] Jones, E.G. & Powell, T.P.S. An electron microscopic study of the laminar pattern and mode of termination of the afferent fibre pathways to the somatic sensory cortex. *Philos. Trans. R. Soc. Lond. (Biol.)*, 1970, *257*, 45-62.

[98] Jordan, M.I. Sequential dependencies and systems with excess degrees of freedom. COINS Technical Report 88-27, Dept. of Computer and Information Science, University of Massachussets, Amherst, MA, 1988.

[99] Jordan, M.I. & Rosenbaum, D.A. Action. In *Foundations of cognitive science*, M.I. Posner, (Ed.), Cambridge, MA: MIT Press, 1989.

[100] Kammen, D. & Yuille, A. Spontaneous symmetry-breaking energy functions and the emergence of orientation selective cortical cells. *Biol. Cybern.*, 1988, *59*, 23-31.

[101] Kelso, S.R., Ganong, A.H., & Brown, T.H. Hebbian synapses in hippocampus. *PNAS* USA, 1986, *83*, 5326-5330.

[102] Kelso, J.A.S. & Stelmach, G.E. Central and peripheral mechanisms in motor control. In *Motor control: issues and trends*, G.E. Stelmach, (Ed.), New York: Academic, 1976.

[103] Kennedy, M.B. Regulation of synaptic transmission in the central nervous system: long-term potentiation. *Cell*, 1989, *59*, 777-787.

[104] Khatib, O. Real-time obstacle avoidance for manipulators and mobile robots. *Int. J. of Robotics Res.*, 1986, *5*, 90-98.

[105] Klapp, S.T. Syllable-dependent pronounciation latencies in number-naming, a replication. *J. Exper. Psychol.*, 1974, *102*, 1138-1140.

[106] Klapp, S.T. & Wyatt, P.E. Motor programming within a sequence of responses. *J. Motor Behavior*, 1976, *8*, 19-26.

[107] Klapp, S.T. Reaction time analysis of programmed control. *Exercise and Sport Science Reviews*, 1977, *5*, 231-253.

[108] Koch, C. & Poggio, T. Biophysics of computation: neurons, synapses, and membranes. In *Synaptic function*, G.E. Edelman, W.F. Gall, & W.M. Cowan, (Eds.), New York: Wiley, 1987.

[109] Koch, C., Poggio, T., & Torre, V. Nonlinear interactions in a dendritic tree: localization, timing, and role in information processing. *PNAS*, 1983, *80*, 2799-2802.

[110] Kohonen, T. *Self-organization and associative memory*. Berlin: Springer, 1984.

[111] Kosslyn, S.M. & Shwartz, S.P. A simulation of visual imagery. *Cognitive Science*, 1977, *1*, 265-295.

[112] Kosslyn, S.M., Pinker, S., Smith, G.E., & Shwartz, S.P. On the demystification of mental imagery. *Behav. Brain Sci.*, 1979, *2*, 535-581.

[113] Kosslyn, S.M. Seeing and imagining in the cerebral hemispheres: a computational approach. *Psych. Rev.*, 1987, *94*, 148-175.

[114] Kubota, K. & Hamada, I. Visual tracking and neuron activity in the post-arcuate area in monkeys. *J. Physiol. (Paris)*, 1978, *74*, 297-312.

[115] Kuperstein, M. Neural model of adaptive hand-eye coordination for single postures. *Science*, March 1988, *239*, 1308-1311.

[116] Lacquaniti, F. Central representations of human limb movement as revealed by studies of drawing and handwriting. *TINS*, 1989, *12*, 287-291.

[117] Lapedes, A. & Farber, R. How neural nets work. In *Neural Information Processing Systems*, D.Z. Anderson, (Ed.), American Institute of Physics: New York, 1988.

[118] Laplane, D., Talairach, J., Meininger, V., Bancaud, J., & Orgogozo, J.M. Clinical consequences of corticectomies involving the supplementary motor area in man. *J. Neurol. Sci.*, 1977, *34*, 310-314.

[119] Latombe, J.C. *Robot Motion Planning*. In preparation.

[120] Le Cun, Y. Une procedure d'apprentissage pour reseau a seuil assymetrique [A learning procedure for an assymetric threshold network]. *Proc. of Cognitiva 85*, Paris 1985, 599-604.

[121] Lee, Y. & Lippmann, R. Practical characteristics of neural network and conventional pattern classifiers on artificial and speech problems. In *Advances in neural information processing systems, vol. 2*, D.S. Touretzsky, (Ed.), San Mateo, CA: Morgan Kaufmann, 1990.

[122] Leinonen, L. & Nyman, G. II. Functional properties of cells in anterolatral part of area 7 associative face area of awake monkeys. *Exp. Brain Res.*, 1979, *34*, 321-333.

[123] Levy, W.B. & Desmond, W.B. The rules of elemental synaptic plasticity. In *Synaptic modification, neuron selectivity, and nervous system organization*, W.B. Levy, J. Anderson, & S. Lehmkuhle, (Eds.), Hillsdale, NJ: Lawrence Erlbaum, 1985.

[124] Linsker, R. Self-organization in a perceptual network. *Computer*, 1988, March, 105-117.

[125] Lockman, J.J. The development of detour ability during infancy. *Child Development*, 1984, *55*, 482-491.

[126] Lozano-Pérez, T. A simple motion-planning algorithm for general robot manipulators. *IEEE J. of Robotics & Automation*, 1987, *RA-3*, 224-238.

[127] Lozano-Pérez, T., Jones, J.L., Mazer, E., O'Donnell, P.A., Grimson, W.L., Tournassoud, P., & Lanusse, A. Handey: a robot system that recognizes, plans, and manipulates. In *Proc. IEEE Conf. on Robotics and Automation*, 1987, 843-849.

[128] Lumelsky, V.J. Effect of kinematics on motion planning for planar robot arms moving amidst unknown obstacles. *IEEE J. of Robotics & Automation*, 1987, *RA-3*, 207-223.

[129] Lynch, J.C. The functional organization of posterior parietal association cortex. *Behav. Brain Sci.*, 1980, *3*, 485-534.

[130] MacKay, D. Hyperacuity and coarse-coding. In preparation.

[131] McCloskey, D.I. Corollary discharges: motor commands and perception. In *Handbook of Physiology, The Nervous System II*, American Physiological Society, 1977.

[132] McNaughton, B.L., Douglas, R.M., & Goddard, G.V. Synaptic enhancement in fascia dentata: cooperativity among co-active afferents. *Brain Res.*, 1978, *157*, 277-293.

[133] Mead, C.A. & Mahowald, M.A. A silicon model of early visual processing. *Neural Networks*, 1988, *1*, 91-97.

[134] Mel, B.W. A connectionist learning model for 3-d mental rotation, zoom, and pan. In *Proc. of 8th Ann. Conf. of the Cognitive Science Soc.*, 1986, 562-571.

[135] Mel, B.W. MURPHY: A robot that learns by doing. In *Neural information processing systems*, p. 544-553, New York: American Institute of Physics, 1988.

[136] Mel, B.W. Further explorations in the learning of visually-guided reaching: making MURPHY smarter. In D.S. *Advances in neural information processing systems, vol 1.*, D.S. Touretsky, (Ed.), Palo Alto: Morgan Kaufman, 1989.

[137] Mel, B.W. & Omohundro, S. Learning with coarse-coded representations. Manuscript in preparation.

[138] Mel, B.W. The sigma-pi column: a model of associative learning in cerebral neocortex. CNS Memo 6, Computation and Neural Systems Program, California Institute of Technology, 1990.

[139] Mel, B.W. & Koch, C. Sigma-pi learning: on radial basis functions and cortical associative learning. In *Advances in neural information processing systems, vol. 2*, D.S. Touretzsky, (Ed.), San Mateo, CA: Morgan Kaufmann, 1990.

[140] Merzenich, M.M & Kaas, J. Principles of organization of sensory-perceptual systems in mammals. In *Progress in psychobiol. and physiol. psychol.*, 1980, *9*, 1-42.

[141] Michalski, R.S., Carbonell, J.G., & Mitchell, T.M., (Eds.), *Machine learning: an artificial intelligence approach*, Vols. I and II, Los Altos: Morgan Kaufmann, 1986.

[142] Miller, K.D., Keller, J.B., & Stryker, M.P. Ocular domninance column development: analysis and simulation. *Science*, 1989, *245*, 605-615.

[143] Miller W.T., Hewes, R.P., Glanz, F.H., & Kraft, L.G. Real-time dynamic control of an industrial manipulator using a neural network based learning controller. Technical Report, Dept. of Electrical and Computer Engineering, University of New Hampshire, 1988.

[144] Minas, S.C. Mental practice of a complex perceptual-motor skill. *J. of Human Movement Studies*, 1978, *4*, 102-107.

[145] Minsky, M. & Papert, S. *Perceptrons*. Cambridge, MA: MIT Press, 1969.

[146] Mishkin, M., Ungerleider, L.G., & Macko, K.A. Object vision and spatial vision: two cortical pathways. *Trends in Neurosciences*, 1988, *6*, 414-417.

[147] Moll, L. & Kuypers, H.G.J.M. Premotor cortical ablations in monkeys: contralateral changes in visually guided reaching behavior. *Science*, 1977, *198*, 317-319.

[148] Moody, J. Fast learning in multi-resolution hierarchies. In *Advances in Neural Information Processing Systems*, D. Touretzky, (Ed.), San Mateo, CA: Morgan Kaufmann, 1989.

[149] Moody, J. & Darken, C. Fast learning in networks of locally-tuned processing units. *Neural Computation*, 1989, *1*, 281-294.

[150] Moody, J. Spontaneous development of modularity in simple cortical models. *Neural Computation*, 1990, in press.

[151] Mountcastle, V.B., Lynch, J.C., Georgopoulos, A., Sakata, H., & Acuna, C. Posterior parietal association cortex of the monkey: command functions for operations within extrapersonal space. *J. Neurophysiol.*, 1975, *38*, 871-908.

[152] Mountcastle, V.B. The world around us: neural command functions for selective attention. The 1975 F.O Schmitt Lecture in Neuroscience, in *Neurosciences Res. Prog.*, Boston, MA.

[153] Mountcastle, V.B., Motter, B.C., Steinmetz, M.A., & Duffy, C.J. Looking and seeing: the visual functions of the parietal lobe. In *Dynamical aspects of neocortical function*, G.M. Edelman, W.E. Gall, & W.M. Cowan, (Eds.), New York: Wiley, 1985.

[154] Newman, W.S. & Hogan, N. High-speed robot control and obstacle avoidance using dynamic potential functions. *Proc. IEEE Conf. on Robotics & Automation*, 1985.

[155] Omohundro, S. Efficient algorithms with neural network behavior. *Complex Systems*, 1987, *1*, 273-347.

[156] Omohundro, S.M. Five balltree construction algorithms. Technical Report TR-89-063, International Computer Science Institute, Berkeley, CA, 1989

[157] Parker, D.B. Learning logic. TR 47, MIT Center for Computational Research in Economics and Management Science, 1985.

[158] Paillard, J., Jordan, P., & Brouchon, M. Visual motion cues in prismatic adaptation: evidence of two separate and additive processes. *Acta Psychol. (Amst.)*, 1981, *48*, 253-270.

[159] Paul, R. *Robot manipulators: mathematics, programming, and control.* Cambridge: MIT Press, 1981.

[160] Piaget, J. The origins of intelligence in children. New York: International Universities Press, 1952.

[161] Pinker, S. Visual cognition: an introduction. In *Visual cognition*, S. Pinker, (Ed.), Cambridge, MA: MIT Press, 1984.

[162] Pinker, S. & Kosslyn, S.M. Theories of mental imagery. In *Imagery: current theory, research, and application*, A.A. Sheikh, (Ed.), New York: Wiley, 1983.

[163] Poggio, T. & Girosi, F. Regularization algorithms for learning that are equivalent to multilayer networks. *Science*, 1990, *247*, 978-982.

[164] Poggio, G. The analysis of stereopsis. *Ann. Rev. Neurosci.*, 1984, *7*, 379-412.

[165] Powell, M.J.D. Radial basis functions for multivariable interpolation: a review. In *IMA Conference on Algorithms for the Approximation of Functions and Data*, RMCS Shrivenham, 1985.

[166] Raibert, M.H. A model for sensorimotor control and learning. *Biol. Cybern.*, 1978, *29*, 29-36.

[167] Rall, W. & Segev, I. Functional possibilities for synapses on dendrites and on dendritic spines. In *Synaptic function*, G.E. Edelman, W.F. Gall, & W.M. Cowan, (Eds.), New York: Wiley, 1987.

[168] Ramón y Cajal, S. Histologie du systeme nerveux de l'homme et des vertébrés. T.2. Paris: Maloine, 1909-1911.

[169] Rawlings, E.I., Rawlings, I.L., Chen, C.S., & Yilk, M.D. The facilitating effects of mental rehearsal in the acquisition of rotary pursuit tracking. *Psychonomic Science*, 1972, *26*, 71-73.

[170] Reeke, G.N.,Jr., Finkel, L.H., Sporns, O., & Edelman, G.M. Synthetic neural modeling: a multilevel approach to the analysis of brain complexity. In *Signal and sense: local and global order in perceptual maps*, G.M. Edelman, W.E. Gall, & W.M. Cowan, (Eds.), New York: Wiley, 1989.

[171] Richmond, B.J. & Wurtz, R.H. Vision during saccadic eye movements. II. A corollary discharge to monkey superior colliclus. *J. Neurophysiol.*, 1980, *43*, 1156-1167.

[172] Ritter, H. & Schulten, K. Extending Kohonen's self-organizing mapping algorithm to learn ballistic movements. In *Neural computers*, R. Eckmiller & C. von der Marlsburg, (Eds.), Heidelberg: Springer, 1987.

[173] Rivest, R.L. & Schapire, R.E. A new approach to unsupervised learning in deterministic environments. In *Proc. of the 4th Int. Workshop on Machine Learning*, 1987, 364-375.

[174] Rizzolatti, G. & Gentilucci, M. Motor and visual-motor functions of the premotor cortex. In *Neurobiology of neocortex*, P. Rakic & W. Singer, (Eds.), John Wiley & Sons, Limited, ©S. Bernhard, Dahlem Konferenzem, 1988.

[175] Roland, P.E., Larsen, B., Lassen, N.A., & Skinhøj, E. Supplementary motor area and other cortical areas in organization of voluntary movements in man. *J. Neurophysiol.*, 1980a, *43*, 118-136.

[176] Roland, P.E., Skinhøj, E, Lassen, N.A, & Larsen, B. Different cortical areas in man in organization of voluntary momements in extrapersonal space. *J. Neurophysiol.*, 1980b, *43*, 137-150.

[177] Roland, P.E. & Friberg, L. Localization of cortical areas activated by thinking. *J. Neurophysiol.*, 1985, *53*, 1219-1243.

[178] Rose, J.E. & Mountcastle, V.B. Touch and kinesthesis. In *Handbook of physiology, Sec. I Neurophysiology*, J. Field, (Ed.), Washington, D.C.: American Physiological Society, 1959.

[179] Rosenblatt, F. *Principles of neurodynamics.* New York: Spartan, 1962.

[180] Rumelhart, D.E. "On learning to do what you want". Talk given at CMU Connectionist Summer School, 1986.

[181] Rumelhart, D.E., Hinton, G.E., & McClelland, J.L. A general framework for parallel distributed processing. In *Parallel distributed processing: explorations in the microstructure of cognition, vol. 1*, D.E. Rumelhart, J.L. McClelland, (Eds.), Cambridge, MA: Bradford, 1986.

[182] Rumelhart, D.E., Hinton, G.E., & Williams, R.J. Learning internal representations by error propagation. In *Parallel distributed processing: explorations in the microstructure of cognition, vol. 1*, D.E. Rumelhart, J.L. McClelland, (Eds.), Cambridge, MA: Bradford, 1986.

[183] Rumelhart, D.E., Smolensky, P., McClelland, J.L., & Hinton, G.E. Schemata and sequential thought processes in PDP models. In *Parallel distributed processing: explorations in the microstructure of cognition, vol. 2*, D.E. Rumelhart, J.L. McClelland, (Eds.), Cambridge, MA: Bradford, 1986.

[184] Rumelhart, D.E. & Zipser, D. Feature discovery by competitive learning. In *Parallel distributed processing: explorations in the microstructure of cognition, vol. 1*, D.E. Rumelhart, J.L. McClelland, (Eds.), Cambridge, MA: Bradford, 1986.

[185] Russell, I.J. Central inhibition of lateral line input in the medulla of the goldfish by neurones which control active body movements. *J. Comp. Physiol.*, 1976, *111*, 335-358.

[186] Schmidt, R.A. *Motor control and learning: a behavioral emphasis. 2nd ed.* pp. 210-220. Champaign, IL: Human Kinetics Publishers, 1988.

[187] Sejnowski, T.J. Open questions about computation in cerebral cortex. In *PDP vol. 2*, J.L. McClelland & D.E. Rumelhart, (Eds.), Bradford: Cambridge, MA, 1986.

[188] Sejnowski, T.J. & Tesauro, G. The Hebb rule for synaptic plasticity: algorithms and implementations. In *Neural models of plasticity*, J.H. Byrne & W. Berry, (Eds.), Cambridge, MA: Academic Press, 1989.

[189] Shepard, R.N. & Cooper, L.A. *Mental images and their transformations.* Cambridge, MA: MIT Press, 1982.

[190] Shepard, R.N. & Metzler, J. Mental rotation of three-dimensional objects. *Science*, 1971, *171*, 701-703.

[191] Shepard, R.N. Psychophysical resonance. In *Neural connections, mental computation*, L. Nadel, (Ed.), Cambridge, MA: MIT Press, 1989.

[192] Shepherd, G.M. & Brayton, R.K. Logic operations are properties of computer-simulated interactions between excitable dendritic spines. *Neuroscience*, 1987, *21*, 151-166.

[193] Shepherd, G.M., Brayton, R.K., Miller, J.P., Segev, I., Rinzel, J., & Rall, W. Signal enhancement in distal cortical dendrites by means of interactions between active dendritic spines. *PNAS*, 1985, *82*, 2192-2195.

[194] Shiller, Z. & Dubrowski, S. Robot path planning with obstacles, actuator, gripper, and payload constraints. *Int. J. Robotics Res.*, 1989, *8*, 3-18.

[195] Singer, R.N. *Motor learning and human performance: an application to motor skills and movement behaviors. Third edition, pp. 425-430.* New York: Macmillan, 1980.

[196] Softky, B. & Kammen, D. Can simple cells learn curves? A Hebbian model in a structured environment. In *Advances in neural information processing systems, vol. 2*, D.S. Touretzsky, (Ed.), San Mateo, CA: Morgan Kaufmann, 1990.

[197] Sperry, R.W. Neural basis of the spontaneous optokinetic response produced by visual neural inversion. *J. Comp. Physiol. Psychol.*, 1950, *43*, 482-489.

[198] Suga, N. Site of attenuation of responses to self-vocalized sounds in echolocating bats. *Science*, 1973, *183*, 1211-1213.

[199] Tawel, R., Eberhardt, S., & Thakoor, A.P. Neural networks for robotic control. Presented at the 1988 Snowbird Meeting on Neural Networks for Computing.

[200] Tesauro, G. & Sejnowski, T.J. A parallel network that learns to play backgammon. *Artificial Intelligence*, 1989, **39**, 357-390.

[201] Trehub, A. Neuronal models for cognitive processes: networks for learning, perception, and imagination. *J. Theor. Biol.*, 1977, *65*, 141-169.

[202] Trevarthen, C. Modes of perceiving and modes of action. In *Psychological modes of perceiving and processing information*, J.H. Pick, (Ed.), New Jersey: Erlbaum, 1978.

[203] Trevarthen, C. Basic patterns of psychogenic change in infancy. In *Regressions in mental development: basic phenomena and theories*, T.G. Bever, (Ed.), New Jersey: Erlbaum, 1982.

[204] Trevarthen, C. How control of movement develops. In *Human motor actions—Bernstein reassessed.*, H.T.A. Whiting (Ed.), Elseview Science Publishers B.V. (North Holland), 1984.

[205] Ullman, S. Visual routines. In *Visual cognition*, S. Pinker, (Ed.), Cambridge, MA: Bradford, 1986.

[206] Van Essen, D.C. Functional organization of primate visual cortex. In *Cerebral Cortex*, Vol. 3, A. Peters & E.G. Jones, (Eds.), Plenum, 1985.

[207] Volper, D.J. & Hampson, S.E. Learning and using specific instances. *Biol. Cybern.*, 1987, *57*, 57-71.

[208] Volper, D.J. & Hampson, S.E. Quadratic function nodes: use, structure, and training. *Neural Networks*, 1990, *3*, 93-108.

[209] Walters, D. Biological vs. artificial representations of variables in fine-grain parallel systems. In *Proc. IEEE Computer Soc. Workshop on Computer Vision*, pp. 298-300, 1987.

[210] Warren, C.W. Global path planning using artificial potential fields. In *Proc. IEEE Conf. on Robotics and Automation*, 1989, 316-321.

[211] White, B.L., Castle, P., & Held, R. Observations on the development of visually-directed reaching. *Child Development*, 1964, *35*, 349-364.

[212] Widrow, G. & Hoff, M.E. Adaptive switching circuits. *Institute of Radio Engineers, Western Electronic Show and Convention, Convention Record, Part 4*, 1960, 96-104.

[213] Willshaw, D.J. & von der Marlsburg, C. How patterned neural connections can be set up by self-organization. *Proc. R. Soc. Lond. B*, 1976, *194*, 431-445.

[214] Wise, S.P. The primate premotor cortex: past, present, and preparatory. *Ann. Rev. Neurosci.*, 1985, *8*, 1-19.

[215] Zipser, D. & Andersen, R.A. A back-propagation programmed network that simulates response preperties of a subset of posterior parietal neurons. *Nature*, 1988, *331*, 679-684.

Index

Accuracy of Reaching, 69
Associative Learning, 9, 14, 123
Back-Propagation, 16, 34
Backtracking, 52, 64
Biological Plausibility, 28, 94, 117
of multi-layer learning, 33
Camera, 10, 5, 18
learning autofocus, 24
Cerebral Blood Flow
relation to reaching, 125, 127
Cerebral Cortex, 9
control of reaching, 121, 124, 134, 136
posterior parietal cortex, 93, 123, 125, 127–128, 131, 133
premotor cortex, 123–124, 133–136
primary motor cortex, 106, 126
supplementary motor cortex, 123–124, 131, 134
Coarse-Coding, 12, 14, 34, 80
example, 13
role in cluster formation, 44
using vision hardware, 7
Collision Detection, 5, 54
definition for MURPHY, 8–9
Conjunctive Encoding, see Multiplicative Connections
Connectionism, 9
representational issues, 12
Connectionist Architecture, 11
diagram, 10
Coordinate Systems
transforms for sensory-motor control, 1, 65
Corollary Discharge, 112–113
Datacube Boards, 7, 14
see also Visual Processing
Decoding, 14–15, 19, 82
see also Coarse-Coding
Development, sensorimotor, 93
Dynamics, 118–119
Ecological Relevance, 3, 86, 98
of reaching, 93, 118
Efference Copy, 112
Extrapersonal Space, 93, 129
cortical representation, 126, 128
Feedback Control, biological, 120
Flailing, 5, 51, 65–66
Free-Space Graph, 83
Function Approximation, see Learning
Gaussian, 7, 12
Generalization, 30, 47
global, 32–33
local, evidence in biology, 99–100
Hand-Velocity Population, 9, 66
Hardware Implementation, 69, 7
Hebbian Learning, 30–31, 35, 37, 43
limitations, 32
locality, 33
Heuristic Search, for motion planning, 54
Hidden Units, 32–33
Higher-Order Unit, 35
relationship to sigma-pi, 41
Intelligent Teacher
doing without, 23
traditional need for, 23
Inverse Kinematics, see Kinematics
Joint Coordinates, planning in, 65
Joint-Velocity Population, 12, 66
Joint-Angle Population, 12, 15, 66
K-d Tree
definition of, 48

for kinematic map, 25
learning with, 37, 48, 72, 74
time needed to construct, 50
Kinematics, 3, 5
biological issues, 117
connectionist representation, 5, 19, 80–81
inverse differential, 17, 66, 68
learning, 20, 66
inverse, 17
learning, 19, 22
multiple inverse solutions, 21, 81
replacing with search, 53
forward, 19, 24, 44
parallelism, 48
traditional methods, 77, 16
difficulties with, 17–18
Learning
by doing, 3, 18, 23–24, 26–27, 77
supervised connectionist paradigm, 26, 28
disadvantages of, 78
forward kinematics, 19
inverse kinematics, 19, 22
inverse differential kinematics, 20
reason for, 1, 18, 77
representational issues, 14
training time, 26, 48, 71, 73
unsupervised, 23, 26
with lookup-table, 27, 35, 37
without intelligent teacher, 23, 26
Local Generalization, see Generalization
Local Minima, 51, 64
Lookup-Table
disadvantages, 39
learning with, 27, 35, 37
neural version, 37–38
Mass-Spring Model, 120
Mental Image, 24
as corollary discharge, 113
biological substrate, 104, 106, 127
cerebral blood-flow studies, 104
for kinematic map, 25
psychology of, 104, 106–107, 102, 108
time to generate, 69
transformations, 105, 107
use in planning, 5, 20, 51, 55, 130
Mental Model
building and using, 24
development in infant, 101
Mental Practice, 5, 108–109, 111–112
Motion Planning, 3, 5, 82
hybrid methods, 87
backtracking, 91
biological substrate, 124
by flailing, 51, 65
cost of degrees of freedom, 92
for multi-link arm, 83, 85–86
geometric algorithms, 5
heuristic search procedure, 54, 64, 90
joint vs. workspace coordinates, 65
with local minima, 51, 64
optimality, 3, 64
quantization of search, 64
size of search space, 64
time to execute, 69–70
using potential fields, 69, 82–83, 88
with obstacles, 53
Motor Control
biological, 120
equilibrium position theory, 126
loading conditions, 120
mass-spring model, 120
Motor Cortex, see Cerebral Cortex
Motor Programming, 106
Multi-Layer Networks, 15
biological plausibility, 33–34
difficulties, 33
more powerful, less biological, 32
Multiplicative Connections, 15–16
use in building receptive fields, 39
MURPHY
comparison to biology, 94, 99
relations to other methods, 77, 79–80, 82, 88, 90
simulated electrophysiology, 129
Muscle Interface, biological, 117
Neural Populations, 5

Parallel vs. Serial Processing, 5, 7, 80, 102
Perceptron, 34
Perceptual Development
infant reaching, 96
learning times, 99, 98
requirements, 95
two modes of reaching, 98
why necessary, 95
Performance
motion planning, 70
optimizing, 69
scaling, 70
learning time, 71
search time, 75
stored data, 74
training time, 73
Physical Setup, 2, 7–8
Posterior Parietal Cortex, see Cerebral Cortex
Potential Fields, 69, 82–83, 88
Premotor Cortex, see Cerebral Cortex
Psychological Issues, 93
summary, 114
Quantization of Search, 64
Radial Basis Functions, see also Receptive Fields, 16, 35
Reaching
cortical substrate, 127
in infancy, 96
two modes of control, 98, 124, 134
with obstacles, 55
Receptive Fields, 12–13, 16, 131
functional form, 12, 14
sensory-motor, 128
Representational Issues, 12, 15
relevance to learning, 14
Robot, 7
Scaling of Algorithms, see Performance
Sculpins, 101
Search Heuristics, 54
Sensorimotor Development, 93
see also Perceptual Development
Sequential Controller, 22
Sigma-Pi Unit, 16, 34
cluster formation, 46
definition, 16, 41
figure, 42
generalization, 47
history, 41
learning rule, 43
relationship to higher order unit, 41
Simulation, on sun workstation, 11
Sparse Coding, 30
Supervised Learning, 26, 28
Supplementary Motor Cortex, see Cerebral Cortex
Synaptic Weights, 5
numbers, 26, 48
number in brain, 75
weakening, 44
System Identification, 27
Thinking, what is, 100
Topographic Maps, 5, 106, 130
Training Time, see Learning
Transfer of Adaptation, 100
Unsupervised Learning, 23, 26
Visual Feedback Control, 65
in humans and animals, 98
Visual-Field Population, 9
Visual Processing
hardware, 7
parallel vision boards, 14
richer role, 5
role in biology, 79
role in robotics, 79, 78, 80
thresholding, 7
unknown camera transform, 3
Visual Representation
binocular, 70
feature extraction, 72
full visual field, 5
of arm, 5, 74, 53
relevance to learning, 72
Visual Routines, 53
Visual-Field Population, 12
Workspace Coordinates, planning in, 65
XOR Problem, 31

Perspectives in Artificial Intelligence

Vol. 1 Kristian J. Hammond, *Case-Based Planning: Viewing Planning as a Memory Task*

Vol. 2 Ray Bareiss, *Exemplar-Based Knowledge Acquisition*

Vol. 3 V. Jagannathan, Rajendra Dodhiawala, and Lawrence S. Baum, editors, *Blackboard Architectures and Applications*

Vol. 4 Thomas R. Gruber, *The Acquisition of Strategic Knowledge*

Vol. 5 Jonathan H. Connell, *Minimalist Mobile Robotics: A Colony-style Architecture for an Artificial Creature*

Vol. 6 Randall D. Beer, *Intelligence as Adaptive Behavior: An Experiment in Computational Neuroethology*

Vol. 7 Bartlett W. Mel, *Connectionist Robot Motion Planning: A Neurally-Inspired Approach to Visually-Guided Reaching*

Vol. 8 James H. Martin, *A Computational Model of Metaphor Interpretation*